SPRINGBOARD KS3 SCIENCE

TEACHER HANDBOOK 1

AF729558

Claudia Allan
Jovita Castelino
Thomas Millichamp
Adam Robbins
Bill Wilkinson

SERIES EDITOR
Adam Boxer

HODDER Education

To order, please visit www.hoddereducation.com or contact Customer Service at education@hachette.co.uk / +44 (0)1235 827827.

ISBN: 978 1 3983 8577 1

First published in 2024 by
Hodder Education
An Hachette UK Company
Carmelite House
50 Victoria Embankment
London EC4Y 0DZ

www.hoddereducation.com

Impression number 10 9 8 7 6 5 4 3 2

Year 2028 2027 2026 2025 2024

Typeset in India

Printed by Hobbs the Printers Ltd, Totton, Hampshire SO40 3WX

A catalogue record for this title is available from the British Library.

CONTENTS

INTRODUCTION

Welcome to Springboard: KS3 Science!

Springboard has been developed by our expert author team with two key goals:

1 to build, develop and sustain students' knowledge and understanding of science
2 to give teachers a powerful resource to support and improve their teaching.

The course is therefore designed not just to be useful or a workload reducer, though Springboard certainly achieves these aims. The main point is to help inform and improve science teaching through evidence- and experience-based practice. In order for students to master new knowledge, teaching needs to involve explicit, narrative-driven explanations, extensive independent practice, constant checking for understanding and a coherent approach to long-term knowledge and retrieval practice. Springboard Science meets all these needs.

This is a revolutionary approach to writing textbooks and schemes of work, and Springboard is groundbreaking in its scope and ambition. Springboard won't just *help* your science teaching, it will *improve* it.

Course elements

The Springboard course is divided first into biology, chemistry and physics and then into units and topics. For example, the B1 unit is about 'Cells' and is divided into smaller topics dealing with microscopes, cell structure, diffusion and so on. Topics are not split into lessons, nor are they of uniform length. You are encouraged to teach the content according to the amount of time it takes, rather than trying to squeeze or stretch it into fixed units of time like a lesson.

While Springboard's content is not structured into fixed units of time, the curriculum plan is designed to enable you to teach through the content over three years. The Practice Books and Teacher Handbooks are organised accordingly, with a pair for each year, each containing multiple units for biology, chemistry and physics. Conceptual progression and interleaving that are built into the resources assume this teaching order is followed.

C1.2 Properties of different states of matter

	Question	Answer
1	Give two properties of gases.	Can flow; can be compressed
2	Give two properties of liquids.	Can flow; cannot be compressed
3	Give two properties of solids.	Cannot flow; cannot be compressed
4	Why can solids not flow?	There are strong forces of attraction between the particles
5	Why can liquids and gases flow?	There are weak forces of attraction between the particles
6	Why can gases be compressed?	There is space between the particles
7	Why can solids and liquids not be compressed?	The particles are already touching

The course is based on the National Curriculum for KS3 and centres around core questions: a delineation and specification of exactly what students are expected to know by the end of the course.

The elements to help deliver the course include:

- a student Knowledge Book, including core questions, key diagrams and worked examples (subject content plus working scientifically)
- student Practice Books, including extensive practice on content just taught and interleaved with topics that have been covered previously
- Teacher Handbooks, including numerous components detailed below
- additional resources on Boost, including class presentations, practical worksheets, careers worksheets, science stories, support sheets, cover work and assessment resources
- online quizzes on Boost for every topic.

For the complete curriculum experience you are encouraged to access Carousel Learning, where question banks containing all the Knowledge Book core questions are available as a flexible quizzing and retrieval resource to all Carousel users with a paid subscription.

Teacher Handbooks

The Teacher Handbooks (THBs) are the centrepiece of Springboard Science and are aimed at supporting you to deliver effective, dynamic, responsive and engaging lessons. They are designed around evidence presented by educational and psychological researchers (e.g. Rosenshine 2012, Clark et al. 2012, Agarwal 2021 and Lemov 2021) as well as the classroom experience of the author team. The THBs broadly follow principles of explicit instruction, an evidence-based approach to structuring teaching and learning.

As such, each THB has several components, broadly following the order:

Topic introduction and objective → General science quiz → Knowledge check →
Guided explanation → Check for understanding → Independent practice and review

Below, the individual components are elaborated on in more detail.

Topic introduction and objective

It is important to have an overview of the topic that is due to be taught, as well as any contextual information about prior knowledge and how it fits within the wider curriculum. Learning objectives are concise, use student-facing language and are simply about the content of the topic, as opposed to any kind of assessment or differentiated approach to lesson planning.

B1.2 Cell structure

Students' learning objective: *I am learning about the basic structure of cells, so I can explain the function of each part.*

Students will have an awareness of what cells are from KS2. They also now have seen cells through their microscope in Topic B1.1.

General science quiz

This is a short quiz testing knowledge from content that has already been taught at some point. The knowledge quizzed here is not generally related to the topic at hand, as the purpose is simply to continually revisit and reconsolidate material that has already been studied. Using retrieval practice like this is an evidence-based strategy that slows forgetting and allows students to steadily accumulate knowledge over time. In most cases, students will complete this quiz silently and on paper, followed by a review.

The questions and answers for the general science quiz are provided in the class presentation (CP) slides on Boost.

Our partnership with Carousel Learning means that you can generate further bespoke retrieval quizzes for your students using the core questions, enabling you to replicate this process every lesson if desired.

Knowledge check

This is a second quiz, aimed at testing the knowledge required for the topic students are about to cover. For example, if students are about to start 'C4.5 Reactions of acids with metals and alkalis', the knowledge check might test their recall of the names of common acids and the pH scale. When students start 'P6 Light', a knowledge check might test their understanding of waves (from P5) and energy (from P1).

The purpose of the knowledge check is to establish whether students are ready to cover the topic in question. If there are holes in their knowledge, they will struggle to access the new content and will fall further behind. Because this purpose relates to assessment, you are urged to perform the Knowledge check using mini-whiteboards (MWBs). If student knowledge is weak or contains gaps, it is *crucial* that you reteach that content there and then, before first rechecking and then moving on to the new topic.

The knowledge check is best conducted one question at a time. As such, the questions are provided on the CP slides, but the answers are not, as it is best to review each question before moving on to the next. Further guidance is provided in the THB if there are any relevant misconceptions or common errors that might need to be identified and tackled. You are encouraged to use strategies like *Show Call* (bringing interesting student responses to the front on a visualiser or some other easy-to-see display) for dissection and further discussion.

Guided explanation

Following a knowledge check, we have provided guidance for, and examples of, how to communicate the relevant content. These are not direct-to-teacher instructions or scripts, but step-by-step ideas and advice that can help you to make your own decisions about how you want to explain a particular concept.

The explanations generally follow a blank-canvas philosophy, which involves starting from a blank board or screen and slowly building a narrative-style explanation, ordinarily involving a diagram. As such, the CP slides often have less on them than you might expect, as a blank slide gives teachers the opportunity to slowly build an explanation and carefully manage students' attention. The THB will then include a suggested 'finished' diagram, with written steps to guide you in constructing the diagram.

The Guided explanation as it appears in the THB

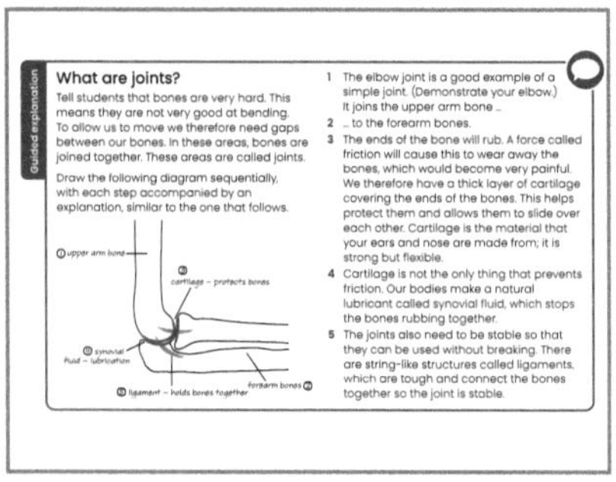

Guided explanation

What are joints?

Tell students that bones are very hard. This means they are not very good at bending. To allow us to move we therefore need gaps between our bones. In these areas, bones are joined together. These areas are called joints.

Draw the following diagram sequentially, with each step accompanied by an explanation, similar to the one that follows.

1 The elbow joint is a good example of a simple joint. (Demonstrate your elbow.) It joins the upper arm bone ...
2 ... to the forearm bones.
3 The ends of the bone will rub. A force called friction will cause this to wear away the bones, which would become very painful. We therefore have a thick layer of cartilage covering the ends of the bones. This helps protect them and allows them to slide over each other. Cartilage is the material that your ears and nose are made from; it is strong but flexible.
4 Cartilage is not the only thing that prevents friction. Our bodies make a natural lubricant called synovial fluid, which stops the bones rubbing together.
5 The joints also need to be stable so that they can be used without breaking. There are string-like structures called ligaments, which are tough and connect the bones together so the joint is stable.

The blank slide that you use in class to build your explanation

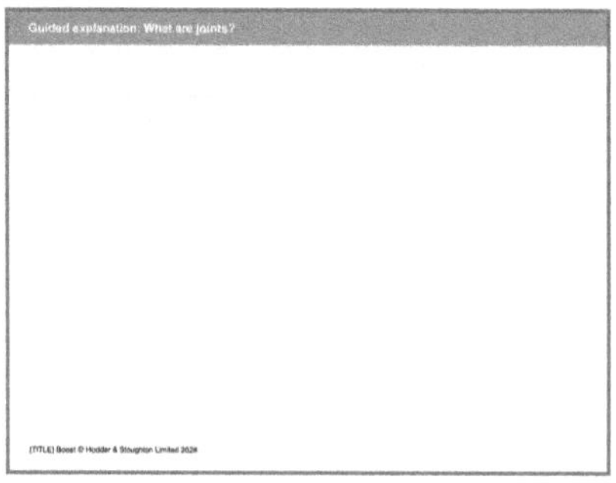

What your finished explanation might look like

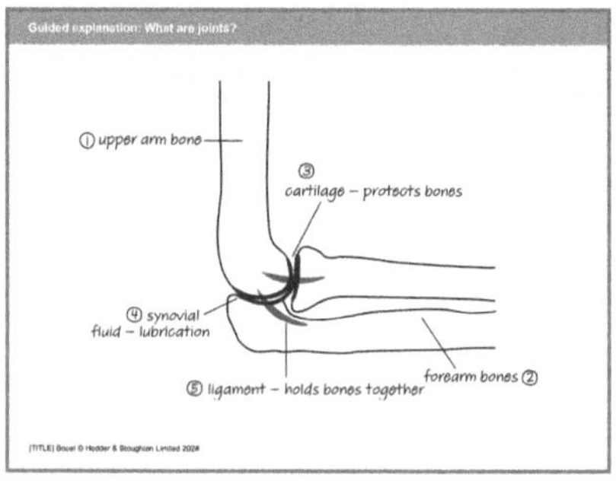

Where diagrams are too complex or detailed to be constructed within a lesson, we have often provided unlabelled outlines or photographs on the CP slides, so you can build your explanation via annotation, highlighting and labelling of the diagram. In these cases, it is a good idea to let students have a thorough look at whatever is on the board first, before calling back their attention to you and working through your explanation.

In terms of the practicalities of delivering explanations like this, schools will vary in terms of the technology and physical resourcing available. Below is a list of accessible methods you can use.

What you might have	How you could use it
Interactive whiteboards	Display the CP slides on the interactive board and write directly onto them.
Projector directly onto a writeable whiteboard	Project the CP slides onto the board and write directly on them.
Writeable whiteboard separate to a projector/screen	Deliver your explanations in a separate location to where you may have a screen display.
Visualiser	Write underneath a visualiser into an exercise book/plain piece of paper/printout of the slides.
Tablet (sometimes called a two-in-one)	Display the CP slides. Use a stylus to write directly onto a computer screen, which is displayed simultaneously to the class via a screen/projector.
Graphics tablet	Display the slides. Plug in the graphics tablet device by USB and write directly onto it. The writing will be recognised and displayed on the computer screen.

When using the CP slides, you may want additional space for more explanations, and in both PowerPoint and Google Slides you can press the letter W in the middle of a presentation to be provided with a brand-new blank slide.

The explanations are always broken down into small steps to avoid overloading students with information.

In general, you may not want to use too many questions during your explanation as this can break up the sense of narrative flow. If you are going to question students during an explanation, using a strategy like *Cold Call* (where you ask the question and then choose a student to respond) is advised.

Check for understanding

Following the explanation, the THB provides suggested questions to ask students before allowing them time to practise independently. It is advised that this is executed on MWBs in a similar way to the knowledge check (as opposed to using individual strategies like *Cold Call*). The goal here is to diagnose and concretise student understanding, so it is important that you get information from everyone in the class, rather than just asking a few students. As with the knowledge check, strategies like *Show Call* are effective at probing and developing student understanding.

It could be that your check for understanding reveals that you might need to go back to your explanation and spend more time over it or try a different approach. As such, the questions we have provided may be extensive, but they cannot be exhaustive, so you are encouraged to be prepared to add to them.

Though the questions will not be provided on the CP slides, you will often find text-based scenarios, unlabelled diagrams or tables to use as an aid when asking them.

Independent practice review

Following a check for understanding, students will be ready to use their independent Practice Books to start consolidating and stretching their knowledge. There are many different styles of question in the Practice Books, including straightforward comprehension/recall, application to unfamiliar scenarios, mathematical calculation/interpretation and specific writing strategies like *Because-But-So*.

The questions are extremely extensive and often intentionally linked to other topics that students have learned previously. This is called **interleaving** and is a deliberate attempt to achieve two things:

1. Make sure all content is revisited on many occasions throughout the course.
2. Help students see the links, relationships, differences and similarities between different areas of the course.

This style of practice may be new to students, so it could be worth starting periods of independent practice by explaining the style and why it is important.

While students practise, you are encouraged to circulate around the room, checking student understanding and offering feedback where appropriate.

Indicative answers to the questions are provided on the CP slides (as well as in PDF form on Boost and the Hodder Education website). The THB notes particularly common mistakes, errors or conflations to look out for when reviewing students' answers. When reviewing student work, placing a student book under a visualiser and using *Show Call* can be highly effective for developing and improving understanding. When asking individuals for their responses, make sure to use strategies like *Cold Call* to ensure that all students are listening and have their voices heard.

Working scientifically

Springboard has taken a revolutionary approach to the working scientifically (WS) content in the KS3 National Curriculum:

- **delineation**: the WS content has been delineated into either core questions or procedures, rather than leave the content as vague or poorly defined statements.
- **sequencing**: this content has then been carefully sequenced and is introduced into the course over time. It is only once students have a base of concrete knowledge that they can begin to encounter and apply the abstract knowledge that constitutes WS.
- **consistent equation calculation approach**: deliberate steps have been taken to slowly introduce the various aspects of completing calculations (e.g. unit conversions, equation selection, rearrangements, multiple steps). These are spread out across the course in a way that builds knowledge without overloading.
- **the EVERY method** for solving equations is consistently used in order to guide students to accurately complete equation calculations. This is one of many different possible models. It is important you have a consistent model for calculations, as it helps reduce the variation within the problems and therefore the cognitive load on the students.

Guided explanation:
How do we use the power equation?

Complete the worked example on the slide, as shown below, to model the EVERY method and demonstrate how we use the power equation. Explain each step line-by-line.

A TV set has a power of 800 watts and is switched on for 30 seconds. Calculate the energy transferred.

Equation	energy = power × time $E = P \times t$
Values	$E = ?$ $P = 800$ W $t = 30$ s
Enter values	$E = 800 \times 30$
Result	$E = 24\,000$
Y(units)	$E = 24\,000$ J

- **practicals**: most units contain at least one core practical, for which student instructions are provided. However, we encourage you to use a *Slow Practical* approach, which does not necessitate using instruction sheets. Instead, it hybridises demonstration with student activity. Broadly, it involves first gathering students around one desk. You then demonstrate a step to the students (often aided by a visualiser), then check students have understood the step (both 'what?' and 'why?') before asking them to complete that step themselves and then return. This allows students to be both 'hands on' and 'minds on', maintaining a purposeful environment where students know what they are doing and why, and you can easily spot and rectify errors. Suggestions for how to apply this method to each core practical are found in dedicated sections in the THB.

Independent practice for the development of practical and WS skills are found throughout the student Practice Books and in student worksheets on Boost. There are two versions of every student practical worksheet on Boost: the full worksheet with details on equipment, safety and method and the *Slow Practical* version, without these details. There are also teacher and technician notes for every practical, which include answers to the independent practice.

A final word: who is it that is Springboarding?

Titles are important and it would be easy to assume that the individuals doing the Springboarding are the students because this course helps prepare and enthuse them for further study. This is true, but only in part, because our goal is not just to Springboard the students, but to Springboard their teachers. We want you to be able to develop and improve your practice in a way that is evidence-based and workload-friendly. We want classrooms to change, to feel different in terms of the methods and strategies teachers are using. We don't want to just slot into your classroom and make things a bit easier. We want to revolutionise it and help you become a better science teacher.

B1 Cells

B1.1 Microscopes

Students' learning objective: *I am learning about microscopes so I can view things too small to see with the naked eye.*

Students may have experience of using magnifying glasses in KS1 and KS2, or even in their everyday experience. By starting with microscopes we can ensure all students have a common experience, which you can continually refer to as they learn about cells.

Knowledge check

As this is the first biology topic, there is no *General science quiz* at the start. Display the *Knowledge check* slide. The questions are best delivered one by one on mini-whiteboards (MWBs). If students do not know the answer to any of them, you should reteach the relevant concept. Answers are not on the slides because it is best to go over them together as a class.

	Question	Answer	Notes
1	What is the smallest thing you have ever seen?	May include: human hair, grain of salt	This question is to get students thinking about small objects and the limits of our eyesight. Answers will vary depending on student experiences.
2	What does the 'naked eye' mean?	Without any lenses or way of making an object appear bigger	Students may also talk about cameras, so ensure that they know these use lenses and sensors.
3	Can you name something so small you can't see it with the naked eye?	May include: germs, particles	Again, this will depend on student experiences.
4	What could you use to see it if you wanted to?	Magnifying glass, microscope, hand lens	

Guided explanation

Why do we need microscopes?

While the idea of microscopes and the experience of using them might be common to students already, this explanation aims to build a common understanding of how a microscope works and to overcome misconceptions around focusing and magnification. Students often see microscopes as devices that just magnify and ignore the concept of focusing. This leads to common errors of language like 'zooming in' that carries over from their experience with digital technology.

Draw the following diagram. Students have not yet been introduced to cells formally, so construct the diagram sequentially, with each step accompanied by a simple explanation, similar to the one that follows.

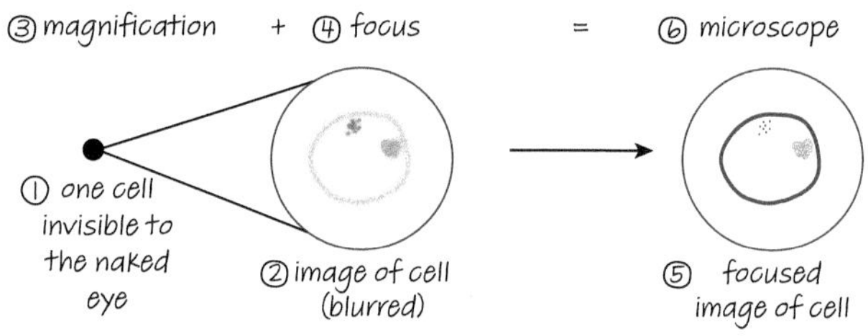

1 A cell is the building block of all living things. Cells are often very small, so are invisible to the naked eye.
2 When we look down a microscope we see what is called an *image* – the original object, just bigger. In this case we want to look at an image of the cell.
3 This is called *magnification* – the image is magnified.
4 The problem we have is that when we enlarge the image it becomes fuzzy – we say it is out of focus. Luckily the microscope is designed to not only magnify but also to focus the image.
5 Once focused we have a clear image of the cell.
6 Overall, we have magnified and focused it. This is what a microscope does: magnification + focus.

The image of all the objects inside the cell now appears big and clear enough to see – the objects are magnified and focused.

Check for understanding

The Check for understanding (CFU) is best conducted using MWBs. Display the *CFU* slide and ask questions like the ones below. Tailor and adapt your questions to the class in front of you.

- What do we call it when we make something look bigger? *Look out for use of the words 'zooming in' instead of magnifying and challenge the use of vague terms like 'make it bigger'.*
- What do we call it when we make a fuzzy, hard-to-see image clearer?
- What two functions does a microscope perform?

The independent practice for this section is included in the questions that follow the guided explanation below.

Guided explanation

How do we use a microscope?

Microscope demonstration

This demonstration is best caried out under a visualiser to ensure that all students can clearly see. If this is unavailable, it might be best to gather students around your desk. You will need a microscope and pre-prepared slide.

1 Demonstrate each part of the microscope and discuss its purpose. Start with the eyepiece and work your way down the microscope to the focus knobs, objective lenses and stage, making sure you are clear about the role of each part.
   - Eyepiece – lens you look down; provides some magnification.
   - Focus knobs – adjust the distance between the slide and objective lens to bring the image into focus.
   - Objective lenses – a group of lenses with different magnifications, which allow you to control how magnified the image is.
   - Stage – where the slide is clipped to, so you can look through it.

2 Get a pre-prepared slide and go through the process of adjusting the microscope to create a focused magnified image.
3 Repeat the process, but this time use questions to get the students to tell you what to do at each stage.

Should you wish to use it, a diagram of a microscope is provided on the slide.

Check for understanding

Display the *CFU* slide and, using MWBs, ask questions like the ones below.

- What is the name of A? *(Repeat for B–G.)*
- What is the function of A? *(Repeat for other parts of the microscope.)*
- What is the purpose of the mirror? *Students may not know the word reflect, so any answer that suggests directing light towards the specimen is acceptable.*
- What parts of the microscope magnify the image? *Check that students say both the eyepiece and objective lens.*

Independent practice review

Students practise independently by answering questions 1–25 from their Practice Books (pages 6–7). Full answers can be found on the *IP answers* slides or in the answer documents on Boost. Notes are only included for questions where useful.

Q	Notes
2	Check to make sure that students move the slide to find some cells before increasing the magnification, as this is commonly missed. Ensure they can say why. Expected answers around it being easier to locate them will suffice.

Core practical

Observing cells using a microscope

A method is provided on the *Practical worksheet* and *Teacher and technician notes* on Boost – including equipment list and safety notes – if you wish to allow your students to proceed independently. However, for most students it will ordinarily be better to take a *Slow Practical* approach following the guidance below.

Students should work in groups of two or three. For each group and for yourself, you will need eye protection, microscope, iodine solution, methylene blue solution (0.1%), tweezers, dental or lolly sticks, paper towel, pipette, pencil, slides and cover slips. You will also need an onion, knife, tiles or cutting boards and some disinfectant (1% Virkon) in a beaker.

1. Assign the equipment by getting students to collect apparatus as needed and return to their workstations. Demonstrate how to correctly carry a microscope.
2. Set up a demo station at the front with the onion and the chopping board.
3. Demonstrate how to peel a single layer of onion cells, stain with iodine and prepare the slide.
4. Set the groups to collect a sliver of onion from the cutting station and mount their own onion slide.
5. Demonstrate the process of using the microscope. This would be a perfect time to use questioning to *Cold Call* students on the parts of the microscope and some of the key features. For example:
   - Why do we only use a thin layer of onion cells?
   - Why do we add the iodine to the onion before viewing under the microscope?
6. When the students have accurately visualised the cells under the microscope, they can record their observations by drawing a simple diagram on the *Practical worksheet*.
7. Repeat the process with a cheek swab. Demonstrate the cheek swab technique and ensure that the students know to place their used dental or lolly sticks in the disinfectant. Again, this is a good chance to check understanding of the parts of the microscope and the purpose of the methylene blue solution.
8. Ask students to complete the independent practice questions on the *Practical worksheet*.

B1.2 Cell structure

Students' learning objective: *I am learning about the basic structure of cells so I can explain the function of each part.*

Students will have an awareness of what cells are from KS2. They also now have seen cells through their microscope in Topic B1.1.

Knowledge check

After working through the *General science quiz*, display the *Knowledge check* slide and, using MWBs, deliver the questions one by one, reteaching concepts as necessary.

	Question	Answer	Notes
1	What do living things need to survive?	Oxygen, food	Look out for answers relating to shelter or protection, which are correct but have no relevance here.
2	When using a microscope, what do we see when we magnify and focus an image?	A bigger and sharper image of the sample	Students may provide specific answers, such as cells, but prompt them to think about how these cells appear when using a microscope and *focusing* the image.
3	Why do we need microscopes to see some things?	Some objects are too small to see with the naked eye	
4	Why did people hundreds of years ago not know about cells?	Cells are too small to see without a microscope, which had not been invented hundreds of years ago	

Guided explanation

What is an organism?

What are the types of organism?

Start by providing a simple definition of the word organism – a living thing. Display the first slide, which shows some examples, e.g. humans, other animals, plants, fungi. Then display the second slide and click to reveal each diagram in turn. Explain that some organisms are made of only one cell and others are made of many, many cells working together.

Draw the following flow chart sequentially, with each step accompanied by an explanation, similar to the one that follows.

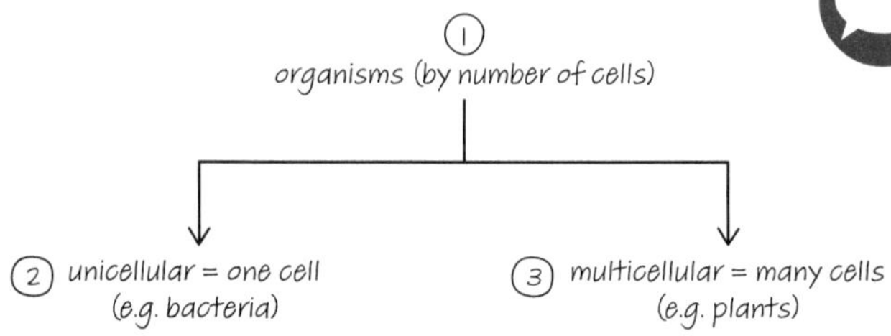

1 Organisms can be one of two types, depending on how many cells they are made of.

2 Some organisms, such as bacteria, are made of only one cell. These are called *unicellular organisms*.

3 Others, such as a *hosta* plant, are made of many, many cells working together. These are called *multicellular organisms*.

What do organisms need to survive?

Build on this by drawing the diagram below sequentially, accompanying each step with an explanation, similar to the one that follows.

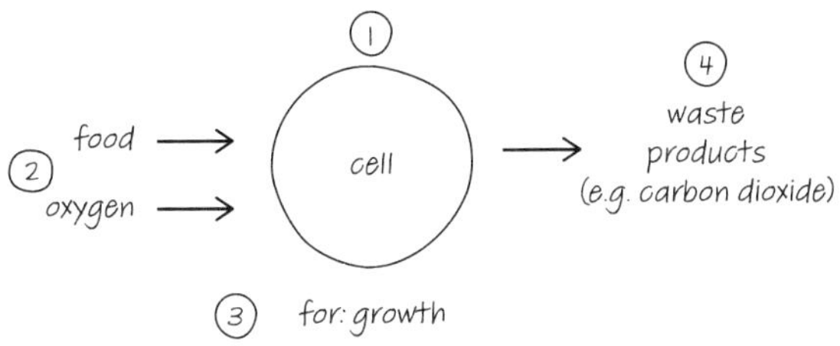

1 All cells within organisms need certain substances to survive.
2 For example, food and oxygen.
3 These also help the cell to grow.
4 Cells also need to get rid of waste products like carbon dioxide.

This will link on to the first sub-cellular structure: cell membrane.

Check for understanding

Display the *CFU* slide and, using MWBs, ask questions like the ones below.

- What are two things cells need to survive?
- What is the difference between a unicellular organism and a multicellular organism?
- Name a waste product that cells need to remove.
- Give an example of a unicellular organism.

The independent practice for this section is included in the questions that follow the guided explanation below.

Guided explanation

What are the basic structures found in cells?

This topic will only focus on four structures found in all cell types. (Note that these structures are found in all cell types, but there are some cells of each type that may not have one or more of these structures.)

Students tend to think that all cells look like a circle or a rectangle, so it is important (with the CFU) that students can identify these structures in any cell.

Draw the following outline of a cell. Add the labels sequentially, with each step accompanied by an explanation, similar to the one that follows.

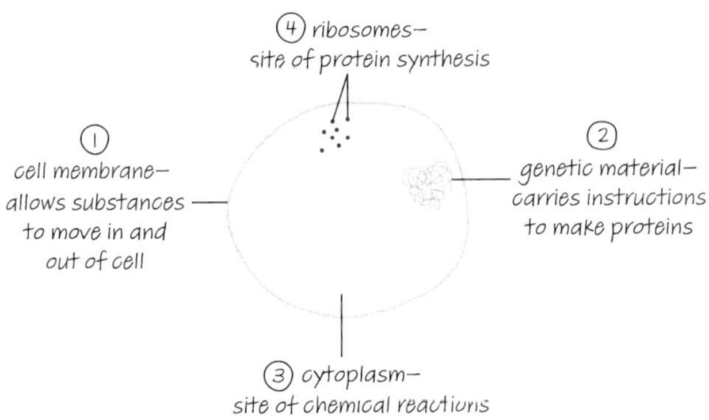

1 Cells need certain parts or structures to help them function well. To allow substances to move in and out of the cell, cells have a part called a *cell membrane*. The cell membrane allows oxygen to move in and carbon dioxide to move out of a cell. All cells have a cell membrane.
2 Cells need proteins to function. The instructions to make these proteins are found in something called *genetic material*. All cells have genetic material in them so they can make proteins.

3 The white space inside the cell is far from empty. It is made of a gel-like material in which all other parts of the cell are suspended. Chemical reactions (needed for cells to survive and function) happen in this part called the *cytoplasm*.

4 The final part found in all cells is the *ribosome*. This is very important because it is where proteins are made.

Say that these four structures are collectively called *sub-cellular structures*.

Check for understanding

Display each of the *CFU* slides in turn and, using MWBs, ask questions like the ones below.

- (First slide) Name the four sub-cellular structures.
- (Second slide) Identify the four sub-cellular structures in these cells. *The aim of this question is to check that all students can identify the structures in cells of different shapes.*

The third slide can act as a visual for the following questions.

- Which sub-cellular structure allows substances to enter and leave a cell?
- Which sub-cellular structure contains the information to make proteins?
- Which sub-cellular structure allows proteins to be made?
- Which sub-cellular structure allows chemical reactions to happen?
- Which of the basic sub-cellular structures in this cell are needed to make proteins? *Genetic material provides the information while ribosomes allow proteins to be made.*
- Cells need oxygen. What does it pass through to enter this cell?

Independent practice review

Student Practice Book questions 26–60 (pages 7–9).

Q	Notes
29	There may be some confusion when asking about the two types of organism, because students may think of the plant and animal kingdoms instead
42	Look out for students assuming that all sub-cellular structures can be seen using a microscope. The microscope that we are using in this unit cannot be used to visualise ribosomes or genetic material, because these structures are too small. The nucleus is visible under a microscope, but this structure will be described in the next topic (B1.3 Animal and plant cells).
44	Look out for students assuming that the shape of the cell is a mistake. It is important for students to understand that cells can vary in shape and size. Part (a) shows a cell with no genetic material. This would not be a mistake if this cell was a red blood cell, but at this stage it is more important to show students that cells generally do contain genetic material (even red blood cells do when immature). When discussing the answer to part (b) emphasise that the cell membrane does have gaps, but that these are very small and only allow specific substances to pass through.

B1.3 Animal and plant cells

Students' learning objective: *I am learning about the general structure of plant and animal cells, so I can identify each type of cell.*

Students should be very familiar with common examples of animals and plants. Now that they understand common sub-cellular structures they are in a good place to learn about the extra ones that plants have.

Knowledge check

After working through the *General science quiz*, display the *Knowledge check* slide and, using MWBs, deliver the questions one by one, reteaching concepts as necessary.

	Question	Answer	Notes
1	What are the four sub-cellular structures found in all cell types?	Ribosomes, genetic material, cell membrane, cytoplasm	
2	What is the function of the cell membrane?	Allows substances to enter and leave the cell	
3	What are the sub-cellular structures needed to make proteins?	Ribosomes and genetic material	Students must think of the instructions *and* the location of where proteins are made, so the answer must include both ribosomes and genetic material.
4	What was the difference in the shape of the onion and cheek cells we viewed in Topic B1.1?	Onion cells tend to be squarer than cheek cells	
5	What is a multicellular organism?	An organism made of many cells	
6	What do plants need to grow?	Plants need sunlight and water to grow	

Guided explanation

What are multicellular organisms?

Introduce the topic by telling students that there are two main examples of multicellular organism: animals and plants. Use the slide to remind them that they have already looked at an example of cells from each of these organisms: onion and cheek cells.

What is a nucleus?

Draw the diagram below, accompanied by an explanation, similar to the one that follows.

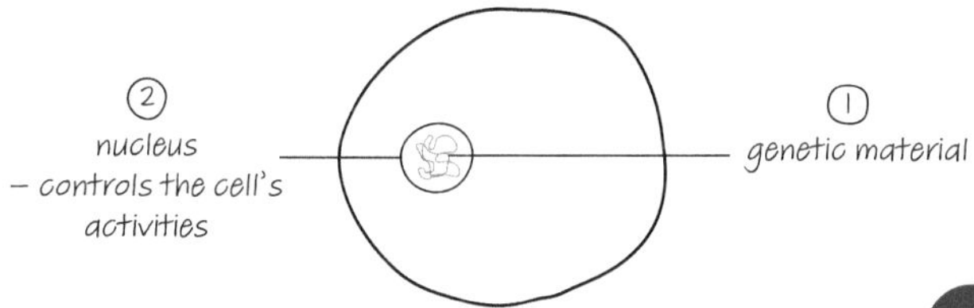

1 Tell students that in multicellular organisms the cells have a membrane that surrounds the genetic material.

2 Explain that this sub-cellular structure is called the *nucleus*. The nucleus controls the cell's activities and it contains the genetic material or DNA.

Watch out for students thinking the nucleus is the brain of the cell. It contains the instructions to make proteins (genetic material) and controls all the activities within the cell. It is not a 'brain' because it does not 'think', and brains are made of cells too.

Check for understanding

Display the *CFU* slide and, using MWBs, ask questions like the ones below.

- What is the function of the nucleus?
- What are multicellular organisms?
- Why is the nucleus not the brain of the cell?
- Where is the genetic material found within a cell?

The independent practice for this section is included in the questions that follow the guided explanation below.

Guided explanation

What are the basic structures found in animal and plant cells?

Draw the outlines of the two cells, as shown below, and label these as an animal and plant cell respectively. Then add the sub-cellular structures and labels sequentially, accompanying each step with student input or explanation, similar to the outline that follows.

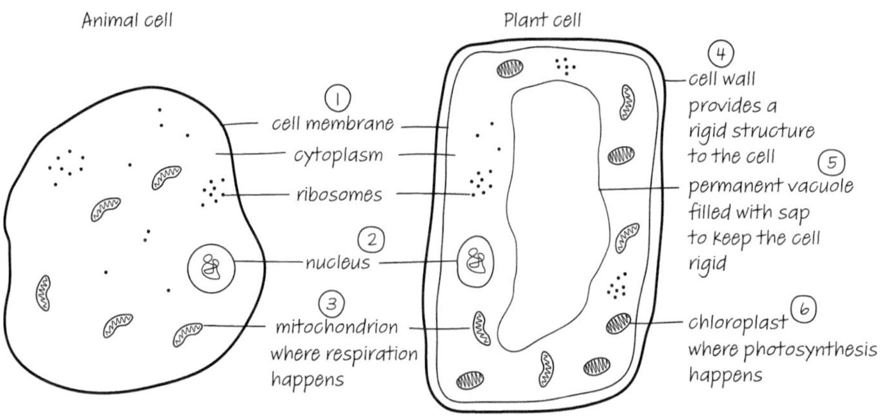

1 Check that students know that both these cells have the four sub-cellular structures. Draw and label the cell membrane, ribosomes, genetic material and cytoplasm as students name these structures.

2 Draw a circle around the genetic material in both cell outlines and check that students can recall this is the nucleus.

3 Introduce the mitochondrion (plural: mitochondria) and say that the process of respiration takes place here. Explain that respiration is a way of releasing energy from food molecules. At this stage, this simplistic definition is sufficient as respiration is covered in Topic B7.1.

Tell students that we have covered all five common sub-cellular structures found in an animal cell. Now you can introduce the three additional sub-cellular structures found in plant cells but not in animal cells.

Explain that these structures allow a plant to perform some different functions to animals.

4 Draw and label the cell wall. Explain that this structure is strong and tough, and that it helps the cell keep its shape.

5 Draw and label the permanent vacuole. Say that plants also have a permanent vacuole to store some of this water in the form of cell sap, which keeps the cell rigid.

6 Draw and label chloroplasts in green, if possible. Explain that plants also need sunlight to grow well. This sunlight is trapped by the chloroplasts and is used in a process called *photosynthesis*. Photosynthesis is a process by which the plant makes its own food (photosynthesis will be covered later in the course). Explain that chloroplasts are green and provide the green colour of plant leaves.

Cells are 3D

Explain the three-dimensional nature of these cells, because students tend to assume cells are 2D in shape. Display the slide and point out the sub-cellular structures on the images so that students can recognise them in different guises.

Check for understanding

Display the first *CFU* slide and, using MWBs, ask questions like the ones below.

- Is this cell an animal or a plant cell?
- Which sub-cellular structure provides structure to the plant cell?
- Which sub-cellular structure is where respiration happens?
- Which sub-cellular structure is where photosynthesis happens?
- Which sub-cellular structure is where cell sap is stored?

Display the second slide and ask questions like the ones below.

- Is this cell an animal or a plant cell?
- How can you tell this is an animal cell?
- Which cell has a cell wall?
- What could these tiny spots be?

Independent practice review

Student Practice Book questions 61–90 (pages 10–12).

Q	Notes
80	This is a good question to check whether students realise humans and sheep are animals, and therefore the cells that make them up will have similar structures.
81	This may be linked to the CFU questions on identifying the type of cell. Emphasise the fact that to identify the type of cell, students must look for certain sub-cellular structures instead of the shape of the cell.
85	It is important to note that some cells are flatter than others, but the key thing here is that students realise that cells do not look exactly like the diagrams we show them. This is why it is important to also show images of cells in 3D and to emphasise that cells actually do have volume (space) and therefore are not 2D.

B1.4 Magnification

Students' learning objective: *I am learning about the magnification equation so I can calculate the image size of a magnified object.*

Now students are aware of the purpose of a microscope and the parts of the cell, we can introduce a mathematical description of magnification. As this is Year 7, we will only focus on one purpose – calculating image size. This should reduce the difficulty and allow all students to master this version before encountering rearrangement later in KS4.

Knowledge check

After working through the *General science quiz*, display the *Knowledge check* slide and, using MWBs, deliver the questions one by one, reteaching concepts as necessary.

	Question	Answer	Notes
1	What is the term for enlarging an image with a microscope?	Magnifying/magnification	Do not accept 'zooming in'.
2	What is the word used for making an image clear enough to see details?	Focusing	
3	Where is the slide placed on the microscope?	Stage	
4	What is the role of the mirror?	To reflect light through the slide	
5	How many millimetres are in 3 cm?	$3 \times 10 = 30$ mm	Students may get confused as to whether they need to multiply or divide by 10. If this is the case, emphasise the fact that there are many millimetres in a centimetre.

Guided explanation

What is magnification?

Recap the idea of magnification from Topic B1.1. Draw the following diagram. As you construct the diagram, accompany it with an explanation, similar to the one that follows.

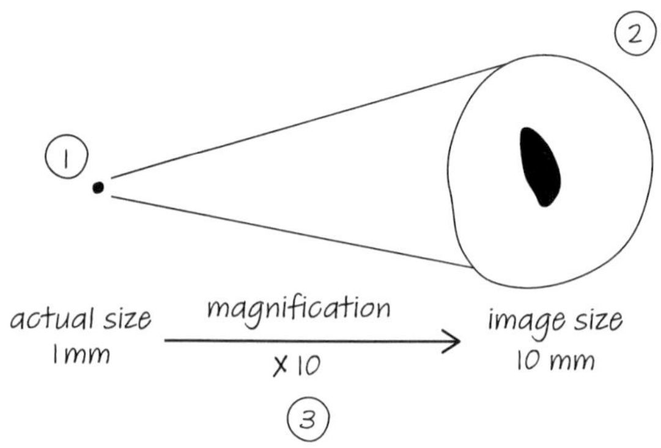

1 Remind students that the actual object is very small.
2 Point out that the image is larger.
3 Then explain that the magnification is the number the actual size needs to be multiplied by to reach the image size.
4 Erase the numbers from the diagram and repeat the process with the following sets of numbers. For the first scenario, add the numbers to your diagram and talk students through the calculation.

	Scenario 1	Scenario 2
Actual size	5mm	1mm
Image size	10mm	20mm
Magnification	×2	?

5 For the second scenario, omit the magnification on the drawing. Ask students what the magnification will be and use questioning to get them to suggest answers, and why. Confirm the answer is ×20 before moving on.

Check for understanding

Display the *CFU* slide and draw the diagram below. Add your choice of numbers for the actual size and magnification.

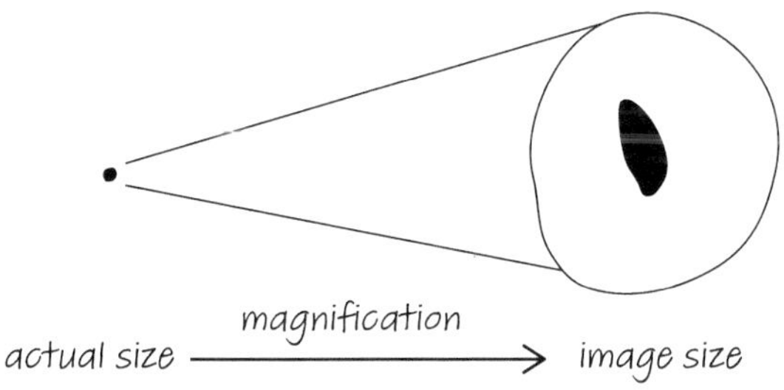

Then, using MWBs, ask questions like the ones below.

- What is the magnification of this microscope?
- Repeat the process with different numbers.

The independent practice for this section is included in the questions that follow the guided explanation below.

Guided explanation

How do we calculate image size?

The following explanation uses the EVERY method for calculations.

What is the magnification equation?

1 Tell students that when we have a relationship between different values, we can use an equation to describe that relationship.
2 Write the magnification equation on the board:

image size = actual size × magnification

3 Emphasise that this equation will work for any object and magnification.

How do we calculate image size?

Complete the following worked example to model the application of the EVERY method to calculate image size, explaining each step line by line. Should you wish to use it, the EVERY template is provided on the slide.

A 0.01mm-long cell is viewed under ×500 magnification. How long will the image be?

Equation	image size = actual size × magnification
Values	a = 0.01mm m = ×500 i = ?
Enter values	0.01 × 500 = ?
Result	5
Y(units)	5mm

Check for understanding

Display each of the *CFU* slides in turn and, using MWBs, ask questions like the ones below.

- Start by asking students to write EVERY down the side of their whiteboard.
- For each calculation, ask them to show their whiteboard to you after each step. *Ensure that you use Show Call to emphasise students who are showing full working.*

Independent practice review

Student Practice Book questions 91–99 (pages 12–13).

Q	Notes
91–99	It is important that you are aware of how many students are getting each question right. You could do this with a simple show of hands after each answer is revealed.

Guided explanation

How do we measure cells?

1 Draw a simple cell and demonstrate how to use a ruler to measure the cell's width. This is easier under a visualiser, but you could also use a whiteboard and metre rule. Be careful to ensure you measure the widest part of the cell.

2 Demonstrate a few examples using cells of different shapes.

3 Then demonstrate how to measure a single sub-cellular structure in a cell.

How do we convert between cm and mm?

Demonstrate this conversion sequentially, as shown below, with each step accompanied by student input or explanation, similar to the outline that follows.

① 10 mm = 1 cm

② 20 mm = 2 cm

③ 300 mm = 30 cm

④ mm → cm (÷ 10); cm → mm (× 10)

1 Start with a simple relationship between mm and cm. If needed, you can project an image of a ruler and count the number of mm in 1 cm, 2 cm, 0.5 cm etc.

2 Add a more complex example and question students. Ask students to explain what they did mathematically to 20 mm to find out the answer. Dismiss answers that talk about doubling the 1 and focus on the correct mechanism.

3 Then change the magnitude of the example. Start by writing the cm first and seeing if students can tell you the mm. Again, get the students to explain how they achieved their answer. You could increase the challenge here by including a decimal if your students find it easy.

4 Finally, summarise the process using a simple diagram.

Check for understanding

Display the *CFU* slide and, using MWBs, ask questions like the ones below.

- Write values and ask the students to show you the converted answers.
- To support students who do not get the right answer first time, it is worth calling on a student who did get the correct answer and asking them what they did to achieve it.

Ensure you have a high success rate on this before moving on.

Independent practice review

Student Practice Book questions 100–03 (pages 13–14).

Q	Notes
100	The answer to this question will vary depending on the size of the page, which is why the answer is not provided. Check the measurements before review.
100–03	It is important that you are aware of how many students are getting each question right. You could do this with a simple show of hands after each answer is revealed.

Guided explanation

How do we calculate total magnification?

Remind students that microscopes do not have just one lens. They have two lenses, and each one magnifies. The first lens magnifies and makes an image and the second one magnifies that image even more.

Display the slide and label the diagram as shown below. Add the labels sequentially, with each step accompanied by an explanation, similar to the one that follows.

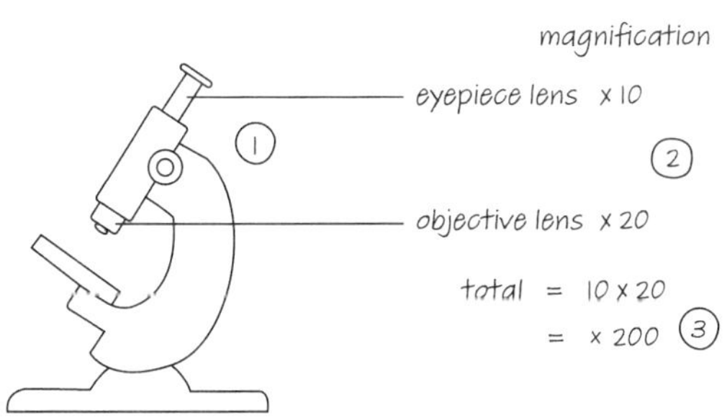

1. Microscopes always have two lenses. What are their names? (Label the eyepiece lens and objective lens as students name them.)
2. In this example, the eyepiece is ×10 and the objective ×20.
3. The total magnification in this example is therefore ×200.

On the second slide, draw the diagram below to show how a small object is magnified twice to give a total magnification of ×100.

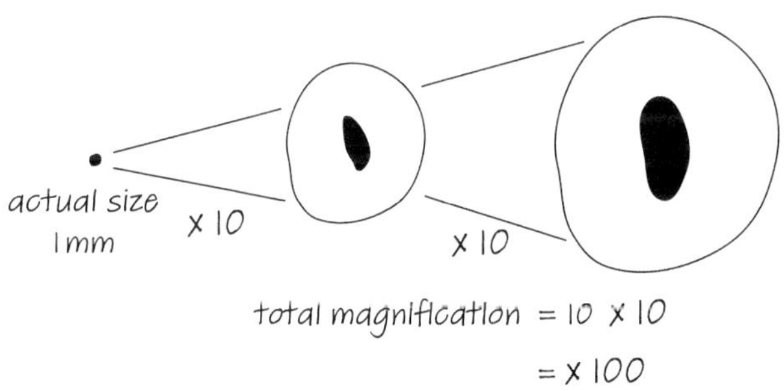

Check for understanding

Display the *CFU* slide and, using MWBs, ask questions like the ones below.

- Add labels to the slide for a new scenario: eyepiece ×20, objective ×50.
- Ask students to write the total magnification on their MWBs so you can see if each student understands.

Independent practice review

Student Practice Book questions 104–17 (pages 14–15).

Q	Notes
117	The final question is much more challenging, so care should be taken when setting it to some students.

B1.5 Unicellular organisms

Students' learning objective: *I am learning about unicellular organisms so I can explain how they are adapted to survive.*

Students will have an everyday understanding of 'germs' and might know words like bacteria and microbes. This topic will allow you to give them a much more detailed understanding of these single-celled organisms, and also provide an opportunity to review everything covered so far.

Knowledge check

After working through the *General science quiz*, display the *Knowledge check* slide and, using MWBs, deliver the questions one by one, reteaching concepts as necessary.

	Question	Answer	Notes
1	What is the function of the cell membrane?	Controls what substances to enter and leave the cell	
2	What is the function of the cytoplasm?	Where chemical reactions happen	
3	What is the function of the ribosome?	Where proteins are made	
4	What are the sub-cellular structures needed to make proteins?	Ribosomes and genetic material	
5	Name two substances all organisms need to survive.	Oxygen and food molecules	Accept 'nutrients'.
6	Give an example of a unicellular organism.	Bacteria, protozoa, unicellular fungi	Accept 'yeast' or other named unicellular organisms. This has not yet been core knowledge, and is the focus of this topic.

Guided explanation

What structures do bacterial cells have?

So far, this unit has looked at how cells are structured in plant and animals, both of which are multicellular organisms. This topic will look at unicellular organisms.

What is the basic structure of a unicellular organism?

1 Recap that unicellular organisms are made of only one cell. Say that examples of these organisms are bacteria.

2 Tell students that bacteria are generally thought of as causing illness, but they also have many uses and benefit organisms. For example, some bacteria live in our gut and can help us absorb important nutrients from our food. Some bacteria are important in the growth of plants and some others are involved in making yoghurt.

3 Explain that as bacteria and other unicellular organisms must survive as just one cell, they look a bit different.

4 Draw the basic structure of a bacterial cell below, without any labels, as an example of a unicellular organism.

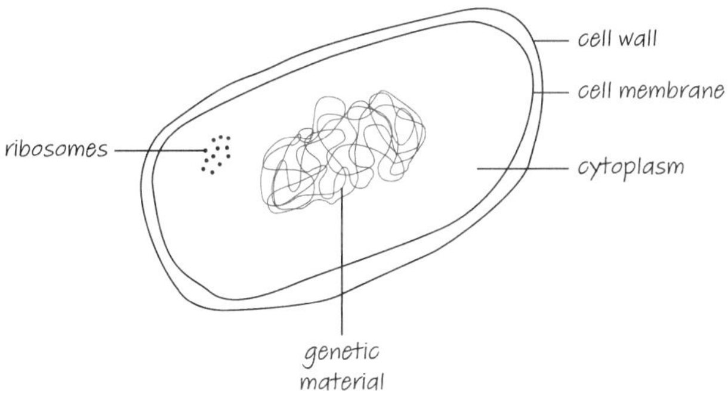

5 Ask students to identify the cell sub-structures from earlier in the unit. Add labels to the diagram as they identify each one.

What additional structures do bacterial cells have?

Explain that bacterial cells need to survive on their own, so they have additional structures that help them with this. Not all bacterial cells have all these additional structures, and some cells have many copies of these structures.

Add further detail and labels to the diagram, as shown below, accompanied by an explanation, similar to the one that follows.

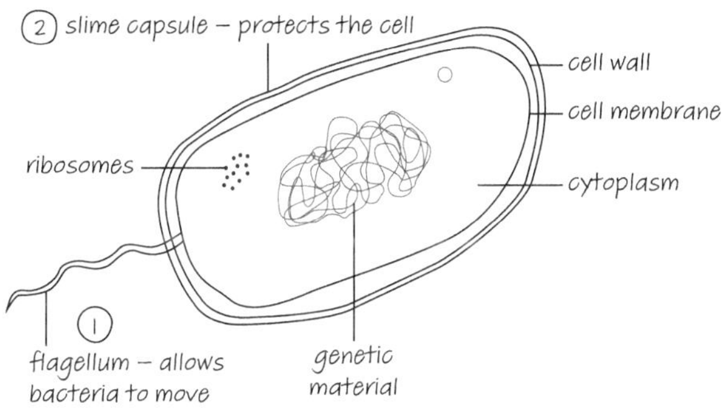

1 Explain that bacteria need some way of moving from one place to another, so they can find sources of food needed for survival. They have a tail-like structure called the *flagellum* (plural: flagella) to help with this.

2 Tell students that while animals and plants protect themselves from disease or damage using many cells, bacteria must rely on themselves so they can have a protective slime capsule around them.

How do these cells compare in size?

It is important to discuss the size of cells at this point.

1 Search online and show students the Scale of the Universe tool. Explain that this tool shows the scale of various objects and that our focus is on organisms and cells, and their relative sizes.

2 Move the scale bar to the left slowly, and pause as below:
- at 10^{-26} to show the largest bacterium
- at $10^{-4.1}$ to show the skin cell
- at $10^{-4.6}$ to show a cell nucleus
- at $10^{-4.7}$ to show a chloroplast
- at $10^{-5.2}$ to show *E. coli*, which is a species of bacterium

3 Explain that bacterial cells do not have a nucleus to surround the genetic material. They also do not have any mitochondria. The reason for this is that bacterial cells are very, very small compared with plant or animal cells. In fact, as we have seen, bacterial cells can be ten times smaller than plant or animal cells. A nucleus would take up too much space and most bacterial cells are about the same size as a single mitochondrion, so the mitochondrion would not be able to fit inside.

Bacterial cells look different

Display the slide showing examples of variations in structure, and explain that bacterial cells do not all look the same.

Check for understanding

Display the *CFU* slide and, using MWBs, ask questions like the ones below.

- Name sub-cellular structures A–E in the cell shown.
- What type of cell is this?
- Why do you think this?

The independent practice for this section is included in the questions that follow the guided explanation below.

Guided explanation

What structures do yeast cells have?

Tell students that bacteria are not the only example of unicellular organisms, and different types of unicellular organism have different structures to allow them to survive.

1 Draw the cell below, without any labels.

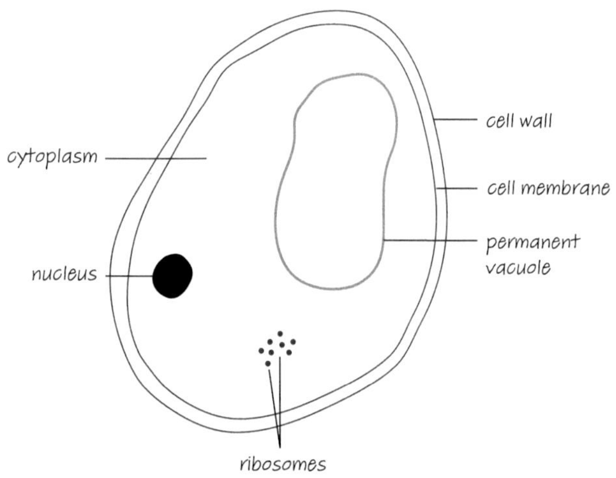

2 Ask students to identify the structures they recognise. They should be able to name all the structures shown. Add labels to the diagram as they identify each structure.

3 Explain that this cell is a typical yeast cell, which is a type of fungus. It has a lot of the structures found in a plant cell but, unlike plants, yeast cannot make their own food using sunlight. This is why they do not have chloroplasts.

Check for understanding

Display the *CFU* slide and, using MWBs, ask questions like the ones below.

- Name sub-cellular structures A–F in the cell shown.
- What type of cell is this?
- Why do you think this?
- How is this cell different from other cell types?

Independent practice review

Student Practice Book questions 118–60 (page 15–19).

Q	Notes
128	Make sure students can make the link between respiration and release of energy from food molecules.
132	Students might consider that bacteria need to move to get oxygen. Guide them to only mention obtaining food molecules through movement. They will learn about how oxygen enters bacterial cells in the next topic.
134	Warn students to read this question carefully because they may think they need to do two calculations here. They only need to calculate in part (a).
138	Technically, photosynthesis is a chemical reaction, but as it needs the specialist structure of the chloroplast, emphasise the difference between the functions of the two structures.
139	Ask students to think carefully about all the components needed for a cell to get energy from food molecules and to make proteins.
145	Ask students to think carefully about whether something is a use or a danger. They have not learned about diseases or the uses of these organisms, but it is a useful exercise for students to evaluate a statement and categorise it.

B1.6 Diffusion

Students' learning objective: *I am learning about diffusion so I can explain how cells transport substances.*

To allow students to understand how substances can be transferred both within and between organisms, they need to understand diffusion. Students will have an understanding of particles from KS2 and possibly from Unit C1.

Knowledge check

After working through the *General science quiz*, display the *Knowledge check* slide and, using MWBs, deliver the questions one by one, reteaching concepts as necessary.

	Question	Answer	Notes
1	Name two substances all organisms need to survive.	Oxygen and food molecules	Allow 'nutrients' for food.
2	What is the function of the cell membrane?	Controls what substances enter and leave the cell	

	Question	Answer	Notes
3	Why is respiration important to cells?	It releases energy from food to keep the cell alive	
4	Name the substances that need to enter a cell.	Oxygen and food molecules	Allow 'nutrients' for food.
5	Name a substance that needs to leave a cell as waste.	Carbon dioxide	Allow CO_2 but not an incorrect molecular formula.

Guided explanation

What is concentration?

From the Knowledge check, students know that all cells obtain important molecules, such as oxygen or food, and they remove waste, such as carbon dioxide. Cells can either obtain or remove substances using a process called *diffusion*. This is incredibly important for unicellular organisms because they are on their own. In multicellular organisms, cells can work together to obtain substances.

Before you discuss diffusion with students, they need to understand the word *concentration*. Draw the diagram below, accompanied by an explanation, similar to the one that follows.

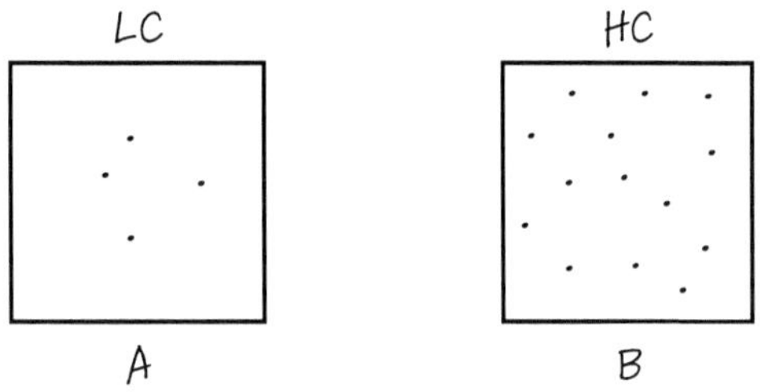

1 Point to Box B and say that this box shows a high concentration of particles. Write HC above the box to represent 'high concentration' (with some classes you might want to write the whole phrase rather than abbreviate).

2 Say that Box A shows a low concentration of particles. Label this box as LC for 'low concentration'.

3 Then draw the diagram below and ask students to identify which side of the rectangle, C or D, shows a high concentration.

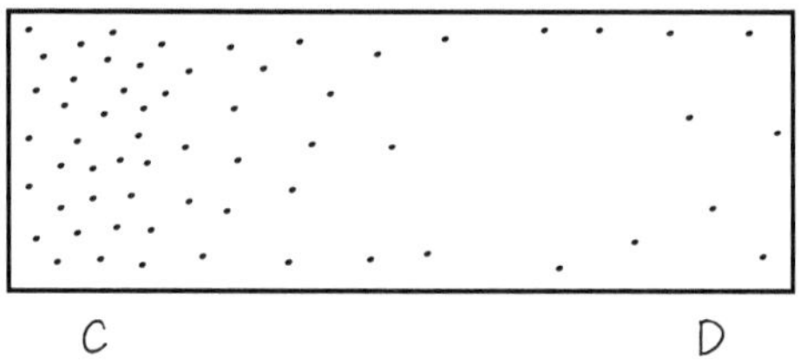

4 Ensure that students have correctly understood that C shows a higher concentration than D.

Check for understanding

Display the *CFU* slide and, using MWBs, ask questions like the ones below.

- Which area shows a high concentration of particles?
- Which area shows a low concentration of particles?

The independent practice for this section is included in the questions that follow the guided explanation below.

Guided explanation

What is diffusion?

Explain that the substances that diffuse might be oxygen, food or waste products, such as carbon dioxide. These particles need to enter and leave a cell and they can do this through the cell membrane of cells. Draw the diagram shown below, accompanied by an explanation, similar to the one that follows.

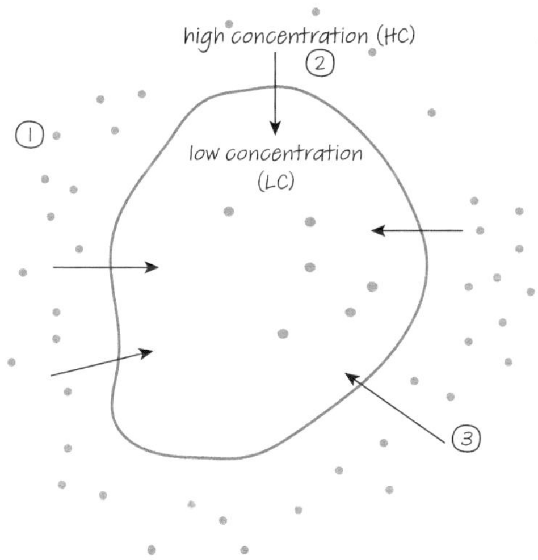

1 Ensure it is clear that there are more particles outside the cell than inside.
2 Use *Cold Call* to ask students which areas are 'HC' and 'LC', to compare the concentration of oxygen outside and within the cell. Add these labels to the diagram.
3 Explain that oxygen will move from the area of high concentration to the area of low concentration. In this example, oxygen will move from outside the cell to within the cell. In other words, oxygen will enter the cell.

Provide a definition for the word *diffusion*, as *the process by which substances move from an area of high concentration to an area of low concentration*. Explain that diffusion happens until the number of particles are equal in both areas; but even then, the particles keep moving.

Check for understanding

Display the *CFU* slide and, using MWBs, ask questions like the ones below.

- Which of the labels, A or B, shows high concentration?
- Draw a copy of the diagram and add arrows to show the direction of movement of particles.
- Define *diffusion*.
- In this case, explain why the substance will leave the cell.
- What kinds of substance enter cells?
- What kinds of substance leave cells?

Independent practice review

Student Practice Book questions 161–71 (pages 19–20).

Q	Notes
161	Accept slight variations of the definition of diffusion. Students must mention areas of high and low concentration.
166	Accept named foods such as sugars, glucose or proteins.

Guided explanation

What factors affect the rate of diffusion?

This topic will cover two ways of increasing the rate of diffusion: surface area and temperature. (Concentration gradient will be covered at KS4.)

Explain that the phrase *rate of diffusion* means *how fast substances move or how fast diffusion happens*. If we say the rate of diffusion is high, this means that diffusion is happening faster.

Draw the diagram below to explain surface area. Construct the diagram sequentially, with each step accompanied by an explanation, similar to the one that follows.

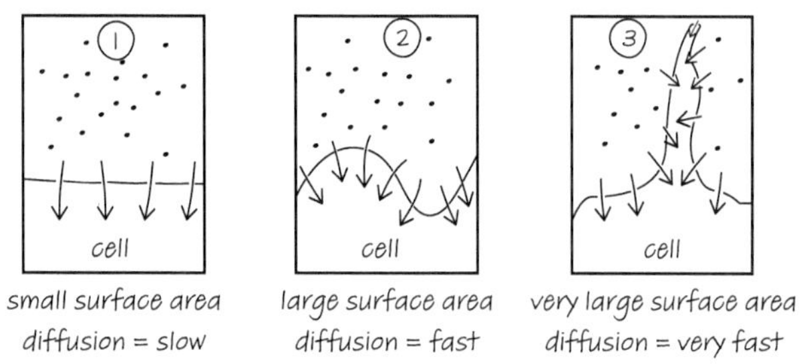

1 Start by just drawing the cell. Explain that the particles will move into the cell through tiny channels, and add these to the diagram.

2 Say that the second cell has a wavy surface, with more channels, so diffusion will be quicker.

3 Say that the third cell has a surface with a big protrusion (bit that sticks out), with even more channels. Explain that this one has a very large surface area, so diffusion will be very fast. The second cell has a large surface, so diffusion is still fast, but the first cell only has a small surface area, so diffusion is slow.

If students need a further physical prompt for surface area, a simple cue of a closed fist being a low surface area and a spread hand being a high surface area might help.

Check for understanding

Display the *CFU* slide and, using MWBs, ask questions like the ones below.

- What factors can affect the rate of diffusion?
- In which way (direction) will particles move in cell A? *(Repeat for B and C.)*
- In which diagram will particles move into the cell?
- Which cell has the highest surface area? How do you know?
- Which cell will have the fastest rate of diffusion? Why?

The independent practice for this section is included in the questions that follow the guided explanation below.

Guided explanation

How does temperature affect the rate of diffusion?

Display the slide with the animation incorporated into the diffusion images. Click through the three stages sequentially, accompanied by an explanation similar to the one that follows.

1 Explain that as we increase the temperature of a substance, the particles within it get more energy. (Students will learn this in more detail in Unit P1.) This makes the particles move around faster.

2 Show the second part of the animation to reveal the thermometer reading being increased and the effect this has on how fast the particles are moving.

3 Show the third part to reveal how increasing the temperature even further increases the speed of the particles.

4 Explain that as particles move faster when temperature increases, the rate of diffusion increases as well. In summary, as temperature increases, the rate of diffusion also increases.

Check for understanding

Display the *CFU* slide and, using MWBs, ask questions like the ones below.

- What factors can affect the rate of diffusion?
- What happens to particles when you increase their temperature?
- In which way (direction) will particles move in cell B?
- In which diagram will particles move into the cell?
- What type of cell is A? *(Repeat for B and C.)*
- Cell C is found in a thermal vent while cell A is found in the Arctic. In which of these two cells will diffusion happen quicker? How do you know?

Independent practice review

Student Practice Book questions 172–213 (pages 20–24).

Q	Notes
176	There is more to this answer because in solids particles do not move as much as they do in liquids and gases. If students have already learned about particles, then this can be included.
179	This question is designed to check whether students understand that once diffusion is complete, the substances do not stop moving. But they maintain an even spread as they move.
187	Diffusion also happens in multicellular organisms and is a vital process, but there are many cells to support each other. As unicellular organisms are only made of one cell, they must rely on diffusion to survive.
198	This diagram shows a typical bacterial cell, but it contains additional structures that have not been covered. Remind students that not all bacterial cells look the same and that there will be additional structures that may not be present in all bacterial cells.

B1.7 Specialised cells

Students' learning objective: *I am learning about specialised cells so I can understand how the body is organised.*

The final topic in this unit looks to build on all that has come so far. Students will be able to name some of the organs from KS1 and KS2 and will have a sound knowledge of the functions of the main ones. We will not cover in detail any particular organ system, as these will come later in KS3, but we will use them as examples.

While specialised cells are beyond the current content of the KS3 National Curriculum, it is important that students have a strong grasp of them so they can better understand the principles of organisation, and how tissues can perform different functions.

Knowledge check

After working through the *General science quiz*, display the *Knowledge check* slide and, using MWBs, deliver the questions one by one, reteaching concepts as necessary.

	Question	Answer	Notes
1	Which sub-cellular structures are found in plant cells but not in animal cells?	Chloroplast, cell wall, permanent vacuole	Animal cells can have vacuoles but they are only temporary.
2	What is the function of the cell membrane?	Controls which substances enter and leave the cell	
3	What is the function of the chloroplasts?	Where photosynthesis takes place	
4	What is diffusion?	The movement of substances from an area of high concentration to an area of low concentration	
5	How does cell surface area affect the rate of diffusion?	The greater the cell surface area, the greater the rate of diffusion	

Guided explanation

How is the red blood cell specialised for its function?

When introducing specialised cells, it is important to make the link between multicellular organisms and the need for each cell to do a particular job. Make sure students are clear on what an adaptation is and, if needed, provide a few examples.

Introduce the idea of specialised cells with the red blood cell (RBC). You will need a lump of plasticine (or similar) for this explanation.

Draw the diagram below sequentially, with each step accompanied by an explanation, similar to the one that follows.

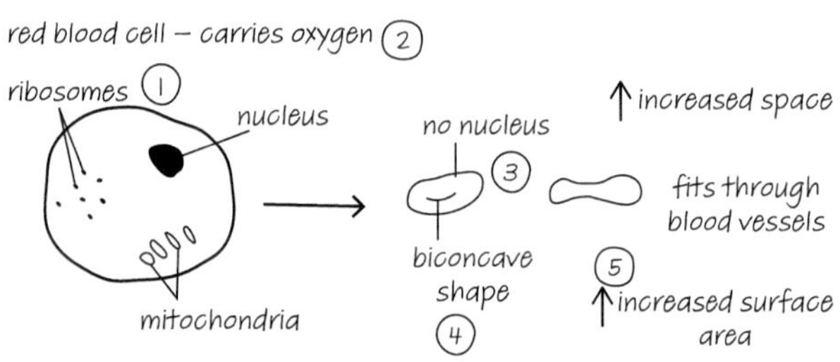

1. Draw a normal animal cell. Use questioning to get the students to name the sub-cellular structures and even ask follow ups on their functions. Add labels to the diagram as they identify each structure.
2. State the RBC function and add this to the diagram with an arrow. Explain that carrying oxygen is an incredibly important job. It is so important that the red blood cell's only job is to carry oxygen. It has no nucleus and is full of a substance called *haemoglobin* that carries the oxygen. Students are not required to know the word haemoglobin right now, so you might choose not to mention it at this stage.
3. Draw the characteristic RBC and discuss that the absence of a nucleus allows for more oxygen to be carried.
4. Introduce the term *biconcave* and define it as 'concave on both sides', which means 'goes inwards'. You can use a lump of plasticine to demonstrate this: roll it into a ball, then squash it down from above and below.
5. Draw the side view of the RBC. Explain how this shape is the best to allow it to fit through small blood vessels. Ask the students to suggest why this is important. Then introduce the increased surface area.

Check for understanding

Display the *CFU* slide and, using MWBs, ask questions like the ones below.

- Why does the red blood cell not have a nucleus?
- What is the function of red blood cells?
- What is the name of a red blood cell's shape?
- Why do red blood cells have a biconcave shape?

The independent practice for this section is included in the questions that follow the guided explanation below.

Guided explanation

How is the muscle cell specialised for its function?

Explain that another specialised animal cell is the muscle cell.

Draw an animal cell similar to the red blood cell example, as shown below. Construct the diagram sequentially, using an explanation similar to the one that follows.

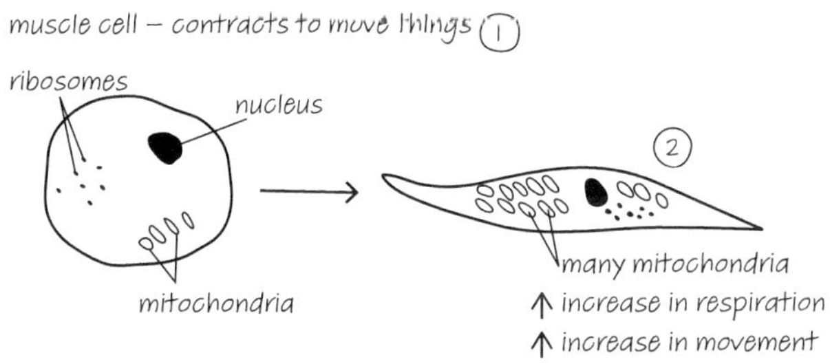

1 The muscle cell's role is to contract to move things.
2 The muscle cell therefore has a longer shape and many mitochondria, because it needs to release lots of energy from mitochondria.

In real life, one muscle cell is not very strong, so many cells work together.

Check for understanding

Display the *CFU* slide and, using MWBs, ask questions like the ones below.

- What is the function of muscle cells?
- What sub-cellular structure do they produce lots of?
- Why do they need so many mitochondria?

Independent practice review

Student Practice Book questions 214–27 (page 25).

Q	Notes
214	Allow the word 'changed' instead of adapted.
217	Do not allow 'increase surface area' for the reason there is no nucleus.
222	You will get some variation in answers here, so it might be worth showing several answers that are different but correct.

Q	Notes
223	Insist on students showing you their working using the EVERY method.
224	*But-Because-So* questions can be challenging and will create a range of valid answers, so allow time for students to think before writing their answer and if they find it difficult consider modelling the first one. Using a visualiser and *Show Call* may be helpful during review.

Guided explanation

How is the root hair cell specialised for its function?

Explain that as plants are multicellular organisms, they also have specialised cells. Tell students that we will now look at two of the many possible examples.

Draw a normal plant cell, as shown on the left below, asking students to help you label the cell, especially the permanent vacuole and chloroplasts. Then add the diagram of the root hair cell, constructed sequentially, using student input or explanation, similar to the outline that follows.

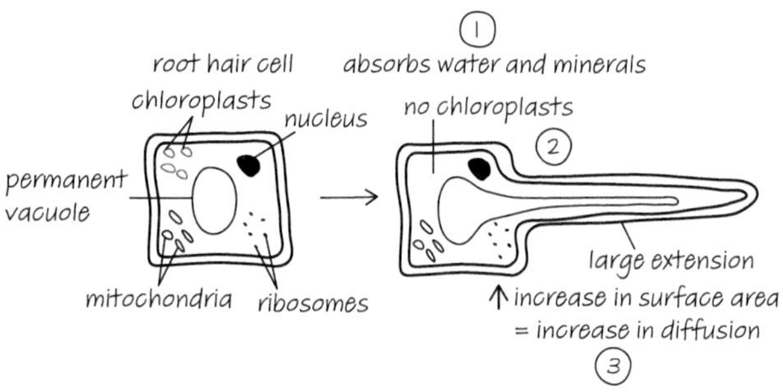

1 The role of the root hair cell is to absorb water and minerals from the soil by diffusion. What factors increase the rate of diffusion?

2 The root hair cell has a long extension or *hair*. This increases the surface area. It also has no chloroplasts because it is underground and so cannot photosynthesise.

3 As the surface area increases, the rate of diffusion increases.

Check for understanding

Display the *CFU* slide and, using MWBs, ask questions like the ones below.

- What is the function of the root hair cell?
- What sub-cellular structure do they not have?
- Why do they not have them?
- How are root hair cells adapted to increase the rate of diffusion?

The independent practice for this section is included in the questions that follow the guided explanation below.

Guided explanation

How is the palisade cell specialised for its function?

Draw the palisade cell below sequentially, with each step accompanied by student input or explanation, similar to the outline that follows.

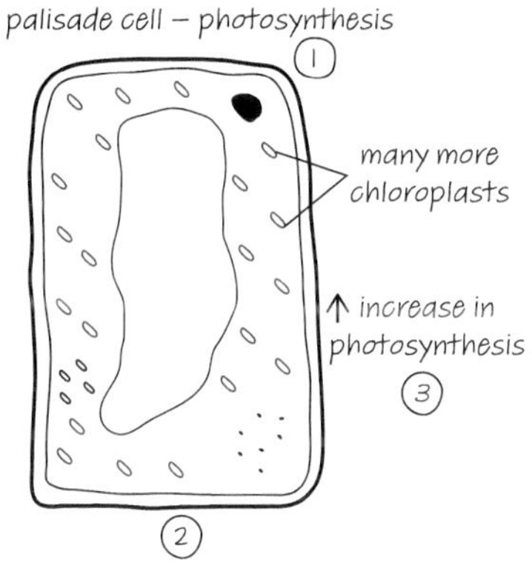

1 The function of palisade cells is to photosynthesise.
2 What sub-cellular structure will they need lots of? (Draw many more chloroplasts, ideally in green.)
3 More chloroplasts will mean more light is absorbed, so faster photosynthesis. We call these cells palisade cells.

Check for understanding

Display the *CFU* slide and, using MWBs, ask questions like the ones below.

- What is the function of palisade cells?
- What sub-cellular structure do they produce lots of?
- Why do they produce so many?
- Why do root hair cells not have any chloroplasts?
- How does the large extension affect the surface area of a root hair cell?
- What is the function of the root hair cell?

Independent practice review

Student Practice Book questions 228–73 (pages 26–29).

Q	Notes
238	Sub-cellular structures can be in any order.
249–51	Must contain comparative language.

B2 Skeletal and muscular systems and organisation

B2.1 The skeleton

Students' learning objective: *I am learning about the skeletal system so I can explain its functions.*

Students will have first-hand experience of their own skeletal system and an awareness of the roles bones and muscles play in movement from Year 3. Some students may have had X-rays or broken bones. The aim of this topic is to exemplify the role of the skeletal and muscular systems before looking deeper into the concept of organ systems in general, introducing the role of nerves in coordination of muscle contraction.

Knowledge check

After working through the *General science quiz*, display the *Knowledge check* slide. The questions are best delivered one by one on mini-whiteboards (MWBs). If students do not know the answer to any of them, you should reteach the relevant concept. Answers are not on the slides because it is best to go over them together as a class.

	Question	Answer	Notes
1	What is our skeleton made of?	Bones	Students should be able to answer this from general knowledge.
2	What do we call the parts of the body where two bones meet?	Joints	You may want to point to a joint on your body like the elbow, so students have a better understanding of what you are referring to.
3	What makes our skeleton move?	Muscles	Accept any reference to tendons and ligaments, but it might be worth exploring how much the students who give these answers know about these, or if it is just a familiarity.

Guided explanation

What is the function of the skeleton?

Introduce the topic by telling students that the human body is made up of 206 bones. The bones all work together, along with our muscles and joints, to allow us to move.

1. Display the image of the skeleton on the slide and give students time to look at it.
2. Call their attention back to you and point out the common bones. Make sure that students are aware that they will not need to memorise the names of the bones. They just need to be aware that they all have very specific names.
3. Make sure students know that the main property of bones is being very hard. This is because they contain calcium.

Add labels to the diagram sequentially, with each step accompanied by an explanation of the function, similar to the ones that follow.

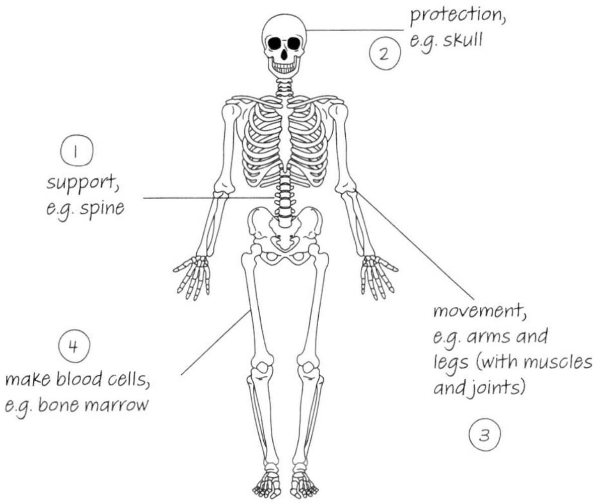

1 The first function of the skeleton is to support the body. For example, your spine supports your body, allowing it to stand upright.

2 The bones also function as protection. Your skull protects your brain and your ribs help protect your lungs.

3 The skeleton also has a key function in movement. Working with the joints and muscles, it allows you to move around freely.

4 The final function of the skeleton is less well known. Your red blood cells do not live for long and need to be replaced regularly. The bones have bone marrow inside them, which allows the bones to make new red and white blood cells.

Check for understanding

The Check for understanding (CFU) is best conducted using MWBs. Display the *CFU* slide and ask questions like the ones below. Tailor and adapt your questions to the class in front of you.

- Give the four functions of the skeleton.
- Which mineral makes bones hard?
- What is the function of the skull?
- What is made in the bone marrow? *Watch out for students saying 'blood', and ensure that they understand it is 'blood cells'.*
- Which function needs the muscles and joints to work with the skeleton?
- The spine is part of the skeleton. What is its function?

Independent practice review

Students practise independently by answering questions 1–10 from their Practice Books (page 30). Full answers can be found on the *IP answers* slides or in the answer documents on Boost. Notes are only included for questions where useful.

Q	Notes
5	Make sure the students say 'blood cells' and not just 'blood'.

Guided explanation

What are joints?

Tell students that bones are very hard. This means they are not very good at bending. To allow us to move we therefore need gaps between our bones. In these areas, bones are joined together. These areas are called *joints*.

Draw the following diagram sequentially, with each step accompanied by an explanation, similar to the one that follows.

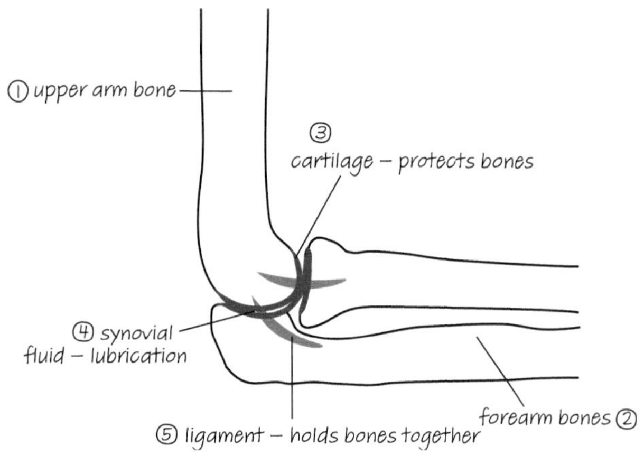

1 The elbow joint is a good example of a simple joint. (Demonstrate your elbow.) It joins the upper arm bone ...

2 ... to the forearm bones.

3 The ends of the bone will rub. A force called *friction* will cause this to wear away the bones, which would become very painful. We therefore have a thick layer of cartilage covering the ends of the bones. This helps protect them and allows them to slide over each other. Cartilage is the material that your ears and nose are made from; it is strong but flexible.

4 Cartilage is not the only thing that prevents friction. Our bodies make a natural lubricant called *synovial fluid*, which stops the bones rubbing together.

5 The joints also need to be stable so that they can be used without breaking. There are string-like structures called *ligaments*, which are tough and connect the bones together so the joint is stable.

Check for understanding

Display the *CFU* slide and, using MWBs, ask questions like the ones below.

- What is the name of the material that covers the ends of the bones?
- What force does synovial fluid reduce?
- What do the ligaments do?
- What would happen if there was no cartilage?
- What three things are found in the joints and what are their functions?

Independent practice review

Student Practice Book questions 11–22 (pages 30–31).

Q	Notes
18	'They would rub together' is all that is required.
20	Accept any answer that states pain in movement or difficulty moving.

B2.2 Biomechanics

Students' learning objective: *I am learning about muscles so I can explain how our bodies move.*

This topic will build on the students' understanding of the skeletal system and introduce the role of muscles. Students will have a good understanding of the adaptations of muscle cells from Unit B1 and the role of muscles in movement from Year 3.

Knowledge check

After working through the *General science quiz*, display the *Knowledge check* slide and, using MWBs, deliver the questions one by one, reteaching concepts as necessary.

	Question	Answer	Notes
1	What works with the skeletal system to help you move?	Muscles	Students might also write joints, ligaments or synovial fluid, but muscles is the target answer.
2	What is the function of muscle cells?	To contract	
3	Why do muscles cells have lots of mitochondria?	So they can release lots of energy from respiration for movement	Challenge the idea of *making* energy.
4	What holds the bones together in a joint?	Ligaments	Make sure students know tendons is the wrong answer here, as this is a common misconception.
5	What is the function of the synovial fluid and cartilage in a joint?	Reduces friction	Allow answers to do with wear and tear and pain.

Guided explanation

What are the main types of muscle?

Introduce the topic by telling students that every muscle cell's main function is to contract. However, different organs need different kinds of muscle cell so they can work as efficiently as possible. There are three main types of muscle, each with its own purpose.

Display the slide and click to reveal each box in turn. Give students time to look at each box before describing the main features. Accompany each step with an explanation similar to the one below.

1 The first type of muscle is skeletal muscle. These are all the muscles that are attached to our skeleton and help us move. We control these consciously, by which we mean as we think about doing something, the muscles move to make it happen. They contract quickly but also get tired easily.
2 The heart is made of muscle. It needs to beat all the time we are alive, even when sleeping. It is made of a different type of muscle cell. Cardiac muscle is branched in structure. We do not have conscious control of cardiac muscle, and it never gets tired.
3 The final type of muscle is smooth muscle. This works slowly, but does not get tired. It lines hollow passages like the intestines, blood vessels and airways to help substances move along. We do not have conscious control of smooth muscle.

Check for understanding

Display the *CFU* slide and, using MWBs, ask questions like the ones below.

- Which type of muscle is connected to the bones?
- Which type of muscle is the heart made of?
- Which type of muscle gets tired?
- Which type of muscle contracts slowly?
- Which types of muscle do not get tired?
- Which type of muscle do we have conscious control of?

Independent practice review

Student Practice Book questions 23–34 (page 31).

Q	Notes
24, 25	Students can struggle with the concept of conscious control. Remind them that they can easily clench their fist but cannot stop their heart beating. Just because we often control muscles subconsciously, it does not mean they are not under conscious control.
34	Students might think they should be talking about the forearm muscles because they do not realise the upper arm muscles are pulling on the forearm to move it. Having them simply feel around their upper arm as they lift their forearm should help them understand what is happening.

Guided explanation

How do muscles help us move?

Explain that muscles transfer energy from respiration in order to contract. This means that they can only pull, and therefore only move bones in one direction. For this reason, they work in pairs – one muscle pulls in one direction and the other pulls in the other direction. These are called *antagonistic muscles*.

Draw the diagram below sequentially, with each step accompanied by an explanation, similar to the one that follows.

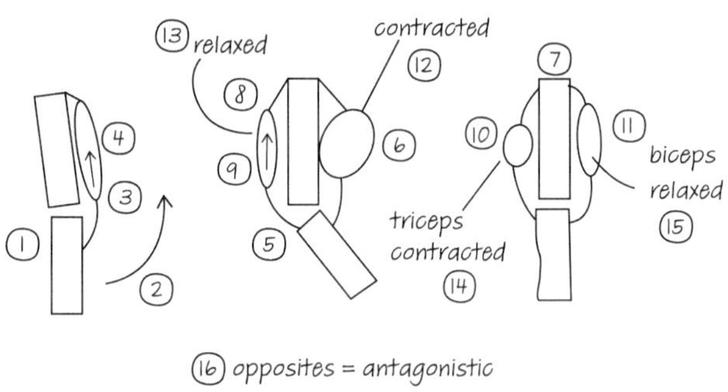

1 Show your arm hanging straight down. (Demonstrate this physically.)
2 I want to lift my arm up like this. (Bend your arm.)
3 I am going to use the muscle at the front of my arm that is connected to the forearm bone here by a tendon.
4 Muscles can only squeeze up together.
5 As my arm bends, the muscle shortens.
6 And now looks like this. Because it is shorter, it has pulled the tendon up, which causes the bone to move up too.
7 Now I want to straighten my arm again. I have a problem. This muscle cannot straighten my arm – muscles can only squeeze up and pull, they cannot push.
8 Fortunately, we have another muscle at the back of the arm. It is also connected to the arm bones by tendons.
9 When this muscle squeezes, it pulls the tendon behind the bone and causes it to go down. The arm is straightened and the front muscle lengthens.
10 This other muscle is now shorter and squeezed.
11 The first muscle now longer. We can see that when one is squeezed the other is not, and when one muscle gets shorter and thicker, the other gets longer and thinner.
12 We call the muscle that is squeezed *contracted*. We say the muscle contracts.
13 The muscle that lengthens we call *relaxed*. It is not pulling.
14 If we look here we can see when the arm is straight the muscle is contracted. This muscle is called the *triceps*.
15 And at the front the opposite is happening. The muscle, called the *biceps*, is relaxed.
16 These pairs of muscles constantly work as opposites: one contracts while the other relaxes. Our word to describe muscles that work in opposition is *antagonistic*.

Check for understanding

Display each of the *CFU* slides in turn and, using MWBs, ask questions like the ones below.

- Which letters show muscles that are contracted?
- Which letters show muscles that are relaxed?
- Which letters show the biceps?
- Which letters show the triceps?
- Which letter is the tendons?
- Which muscle is contracting when the forearm moves up?
- Which muscle is contracting when the forearm moves down? *(Repeat for relaxing.)*
- What do we call muscles that work in pairs to move a limb?
- True or false: muscles can push? Explain your answer.

Independent practice review

Student Practice Book questions 35–51 (pages 31–32).

Q	Notes
38	Make sure students do not mention creation of energy by respiration, but the releasing of energy by or from respiration.

Core practical

Testing muscle strength

A method is provided on the *Practical worksheet* and *Teacher and technician notes* on Boost – including equipment list and safety notes.

This practical is best done with a few volunteers at the front of the class initially. The best equipment to use is a grip strength dynamometer (if your department does not have one, then check with your school's PE department) and newton scales (scales you stand on, but which have N not kg).

1 Test the grip and leg strength of the volunteers, recording the results in the table provided on the slide.
2 Once you have demonstrated how to measure the two types of strength, ask students to work in pairs to record their own strengths in a similar table on the *Practical worksheet*.
3 Ask students to complete the independent practice questions on the *Practical worksheet*, which are based on this practical and the unit content so far.
4 It is recommended that the conclusion for this practical be very superficial. Simply ensuring students understand that their legs are stronger than their forearms and linking this to muscle size is sufficient.

Guided explanation

How can we compare people's strength using graphs?

This is the first time students will have come across the concept of graphing at KS3. Below is a possible route to help introduce them to why and how we use graphs.

1 Display the first slide and explain that a group of students are taking part in an experiment. They all had their shoulder strength measured in a similar way to you in the science lab and then went out on to the field. They were then asked to throw a heavy metal ball, a bit like a cannon ball, called a

a shot put. The distance each student managed to throw the ball was measured and recorded.

2 Display the second slide and say that these are their results. Let students look at the table for themselves first then call their attention back to you.

3 Explain that by just looking at the table, it is hard to notice a trend (pattern). The numbers go up and down and it is hard to see who is the strongest, who is the weakest and who can throw the ball the furthest. To show the trend, or relationship, between the two things we are investigating, we need to draw a graph. A graph shows the relationship between things visually, which makes it easier to see.

Note: This explanation does not use the concept of variables, which are first taught in Topic P2.2, and some schools may not have covered that yet. If you have covered variables, then make sure to first do a knowledge check on variables, and then include the terminology in your explanation. If you are yet to cover variables, use this simplified version. Similarly, there is more introductory graphing work in Topic P2.2, so if you have already taught that topic, you may be able to move faster here.

Tell students you are going to show them how to draw a line graph. Demonstrate this by first drawing the axes below, accompanied by an explanation, similar to the one that follows.

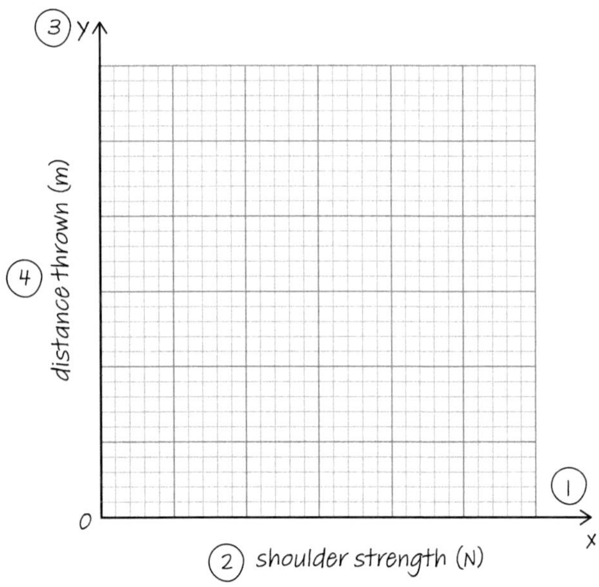

1 This is the *x*-axis. An X is a 'cross', and the *x*-axis goes *across* the paper horizontally.

2 The *x*-axis is where we will put the thing we changed. In this case we looked at people with different shoulder strength. It will always be the first column of numbers.

3 Another line goes vertically. This is called the *y-axis*. We can remember this by telling ourselves 'Y goes high'.

4 The *y*-axis is where we will put the thing we measured. In our case it was the distance thrown. It will always be the second column of numbers.

Point out that there is no place for names. Explain that in this graph we are not going to worry too much about names, as we are more interested in strength than the person throwing the ball.

Plot points and draw a line of best fit, as shown below. Do this sequentially, with each step accompanied by an explanation, similar to the one that follows.

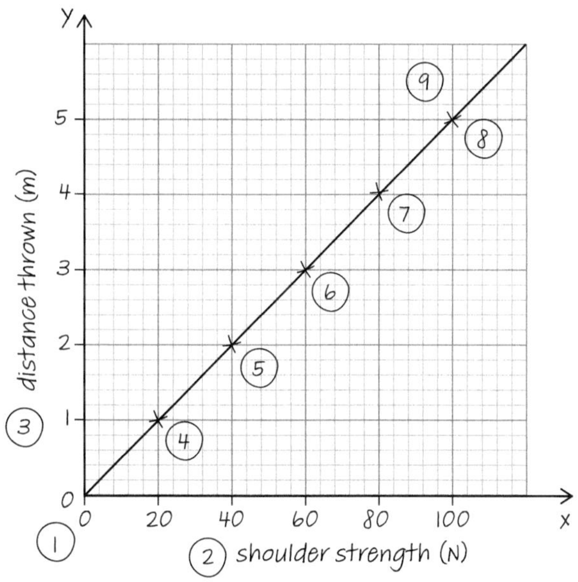

1 The point where the *x*- and *y*-axis meet is called the *origin*, and it has a value of zero.

2 We now need to write the scale for the *x*-axis. The scale contains all the values we need on a number line. Crucially the numbers must be equally spaced so the distance between two numbers is the same. For example, in this graph we have five small squares between 0 and 20 and five small squares between all the other values, because they go up in 20s.

3 For the *y*-axis the scale is different. The numbers only go up in 1s, so we have five small squares between 0 and 1 and the other numbers up to 5. To plot these points, we might need to work out what each small square is. To do this, we work out the difference between two lines on the scale and divide by the number of small squares between them. For example, on the *x*-axis we have two lines that have a difference of 20. There are five small squares between them. Therefore, dividing 20 by 5 tells us that each square has a value of 4 N.

4 Now it is time to plot the first point. We remember the phrase 'Along the corridor, up the stairs' to remind ourselves that we go along the *x*-axis and then up the *y*-axis. So our first point is 20 N and 1 m, so we go along to 20 and up 1 and draw an X.

5 We do the same for 40 N and 2 m.

6 We repeat this for the next three points.

9 Once we have all the points, we look at them to see if there is a trend. In this case, there is a nice straight line from the origin. We therefore take our ruler and carefully draw the line that goes as close to each point as possible. This is called the *line of best fit.*

Students practise independently by completing the *Drawing graphs worksheet* on Boost.

Check for understanding

Display the first *CFU* slide and, using MWBs, ask questions like the ones below.

- On this scale, what would one small square be worth?
- If this was on a graph would it be the *x*- or *y*-axis?
- Put an X somewhere on the scale and ask students what point it represents; repeat as necessary.
- Sketch this scale and mark where you would expect 5.5 to be.

Display the second, third and fourth slides and ask questions like the ones below.

- On this graph I have plotted the points in the table but deleted the scales. What would each mark on the *x*-axis be?
- What would one square on the *x*-axis be?
- What would each mark on the *y*-axis be?
- What would one square on the *y*-axis be?

Display the fifth slide and ask questions like the ones below.

- Here the scale is included. What would be the value of the cross marked 1 on the *x*-axis?
- What would be its value on the *y*-axis?
- What is the value of the point marked 2 on the *x*-axis?
- What value on the *y*-axis is the point marked as 2?

Display the sixth slide and ask questions like the ones below.

- What is the *x*-axis value of A? (*Repeat for B–D.*)
- What is the *y*-axis value of A? (*Repeat for B–D.*)

How do we know what line of best fit to draw?

Display each of the slides in turn, which show a series of pre-drawn graphs for you to draw 'correct' or 'incorrect' lines of best fit on. Work through the graphs sequentially, accompanying each one with student input or explanation, similar to the outlines given below. In each case, respond to the needs of your class, and use more lines and graphs if needed.

Display the first slide, which shows an example of a perfect linear relationship.

1 Click to show students the version with the line of best fit completely wrong. Ask why it is wrong.
2 Ensure students are clear about where the line of best fit should be drawn. Click to reveal this.

Display the second slide.

1 Tell students that some data have been plotted on this graph, and point out that the data are a little spread out.
2 Say that we need to draw a line of best fit. This line needs to go close to all the points. A good line of best fit has an equal number of points above and below the line.
3 Add the line of best fit. Ask students if this is a good line of best fit. Make sure they understand that it is not because it touches some points, but it is not near all the points.

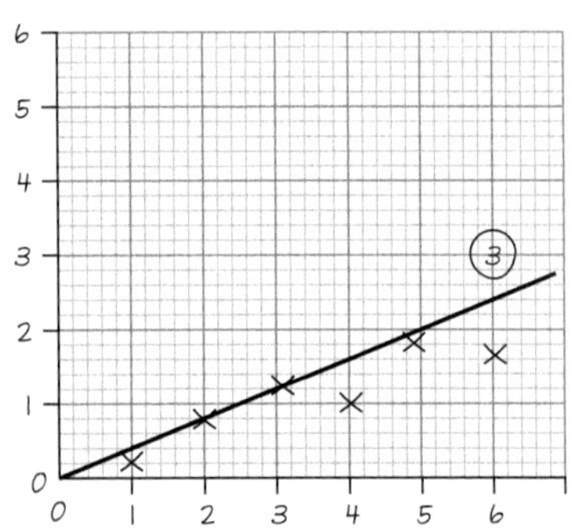

4 Click to show a second copy of the graph. Add the line of best fit. Point out that this line of best fit is close to all the points even though it does not touch any points. It has the same number of points above and below the line.

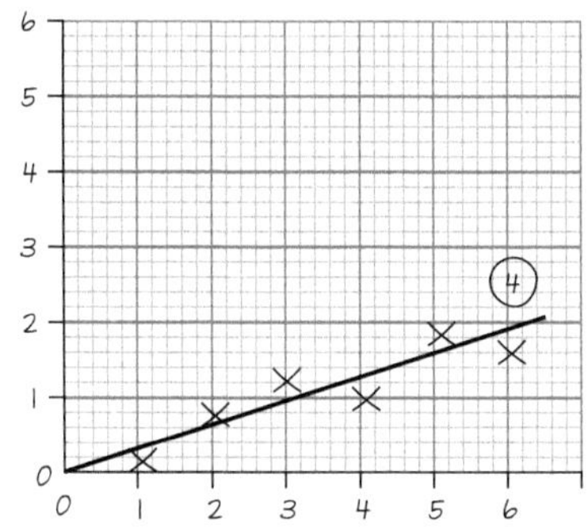

Display the third slide.

1 Add the line of best fit (below). Ask students what they think of the line this time. It goes near all the points, but it is not a straight line.

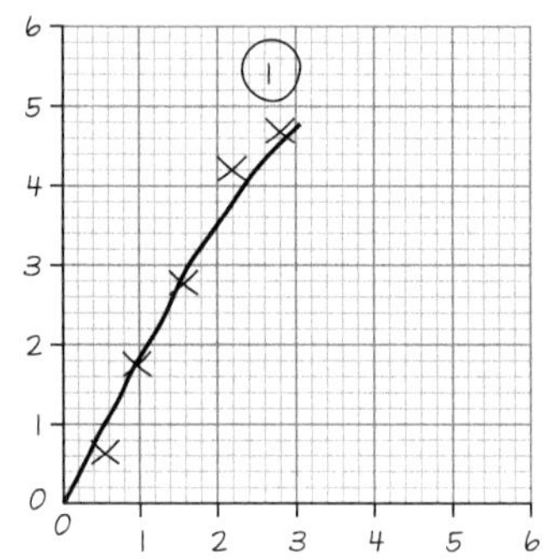

2 Click to show a second copy of the graph. Add the line of best first (below). Say that this line is better. Some points are on the line and others are close to the line, and it is straight.

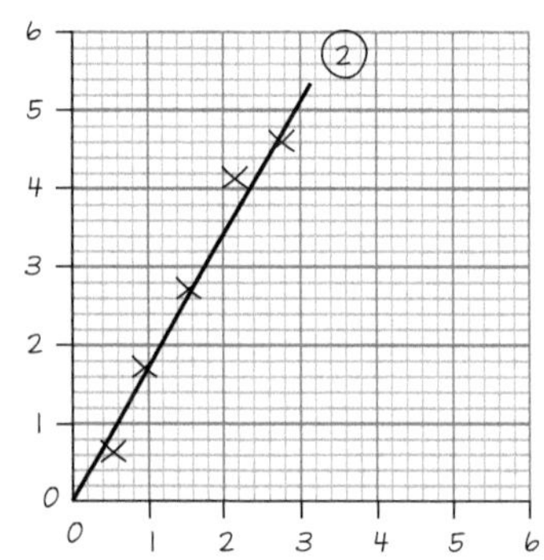

Display the fourth slide and add each line of best fit shown below in turn.

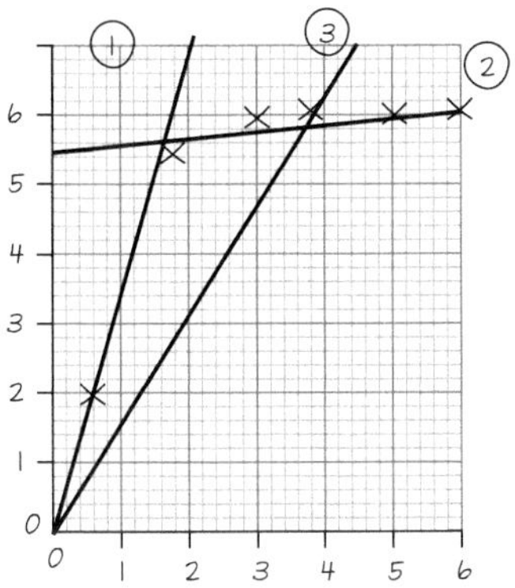

1. Point out that this straight line only hits the first few points.
2. This one only hits the last few points.
3. This one technically has equal points on either side, but it is not close to most of them and so is also wrong.

Display the fifth slide and draw the two lines in turn.

4. Ask if we should just draw it like this. *No, definitely not.*

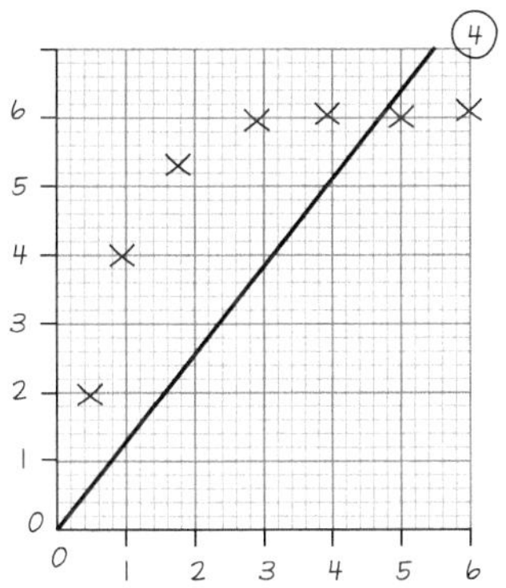

5. Explain that sometimes our line of best fit is a curve. In this case, a curve hits all the points. Importantly, the curve will be a simple curve with only one bend in. It will not snake all over the place.

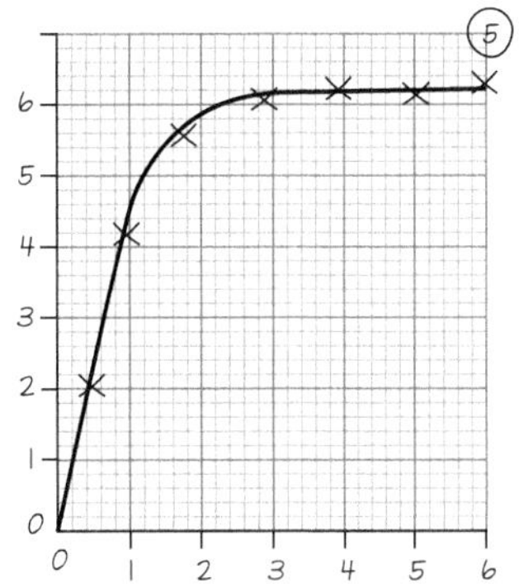

Finally, display the sixth slide and add each line of best fit in turn.

1. Point out that the same goes for this example – we do not draw straight lines here.

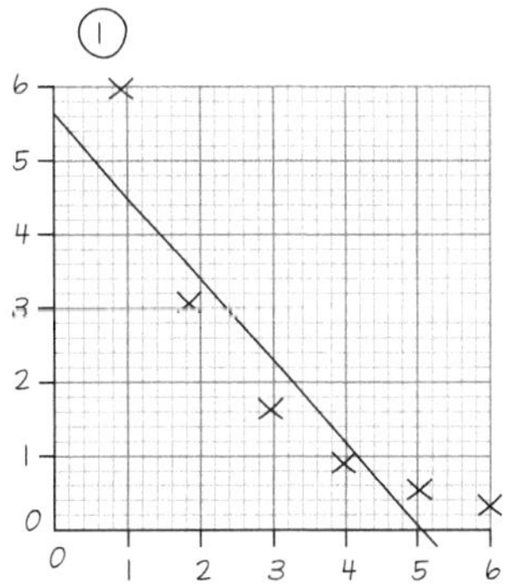

2. Instead, we draw a curve like this.

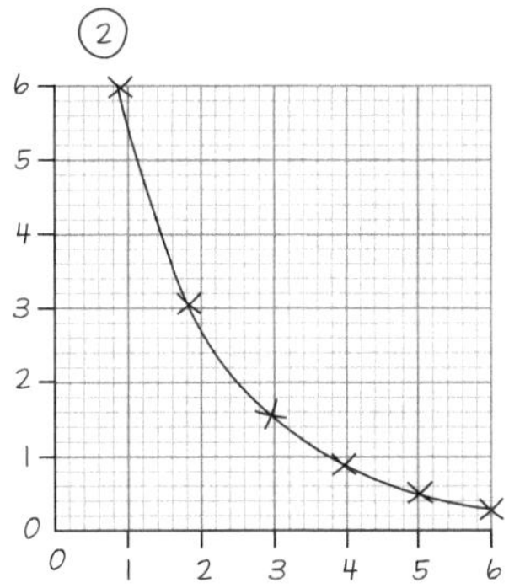

Students practise independently by answering the questions on the *Drawing lines of best fit worksheet* on Boost.

Check for understanding

Display each of the *CFU* slides in turn and, using MWBs, ask questions like the ones below.

- Ask students to sketch the line of best fit. If students are struggling to identify this correctly, create more examples to help them gain more practice.

Independent practice review

Student Practice Book questions 52–58 (pages 33–34).

Q	Notes
54	It should not go through the origin.
55	It must touch the first point and the y-axis.
56	It does not need to touch the origin
58	Students struggle to draw curved lines. Their curve must be a smooth line that is not sketched or joined dot to dot

B2.3 Principles of organisation

Students' learning objective: *I am learning about organisation so I can explain how the body's cells work together to form our organs.*

This topic provides a strategic review of the specialised cells from Unit B1 and ties it into the concept of organs and organ systems. Students should be able to name some common organs and their basic functions from prior experiences, and this can be built on, although the digestive system will be specifically taught later in the course.

Knowledge check

After working through the *General science quiz*, display the *Knowledge check* slide and, using MWBs, deliver the questions one by one, reteaching concepts as necessary.

	Question	Answer	Notes
1	Name two specialised animal cells.	Muscle cell, red blood cell	Many other options are available, but these two were specifically mentioned in Unit B1.
2	Give one adaptation of the muscle cell.	Lots of mitochondria	Make sure students who say 'to contract' understand that is a function and not an adaptation.
3	What is the function of red blood cells?	To carry oxygen around the body	Not to carry waste away – that is the function of the plasma.
4	What is the function of the muscular-skeletal system?	To allow the body to move	
5	What is a multicellular organism?	An organism made of many cells	

Guided explanation

How is the nerve cell specialised for its function?

Draw the diagram below sequentially, with each step accompanied by an explanation, similar to the one that follows.

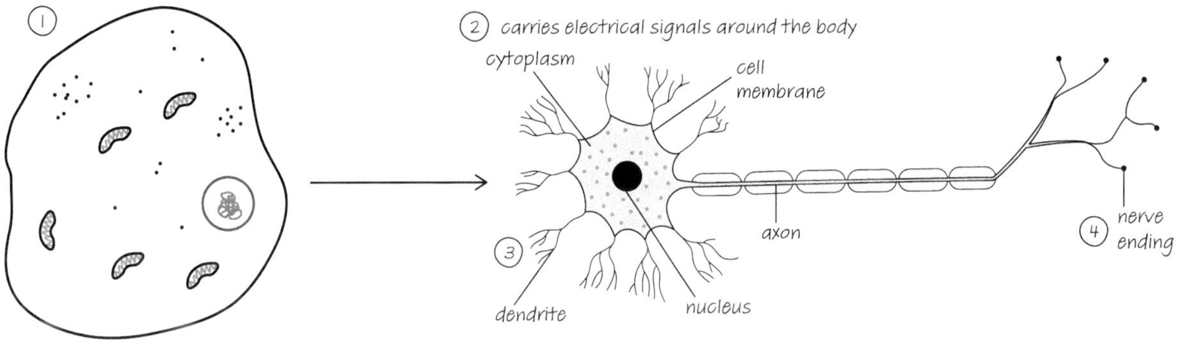

1 A non-specialised animal cell is very familiar to us – it has its cell membrane, cytoplasm, mitochondria, nucleus and ribosomes.
2 Nerve cells are very different because they have a special function. They carry electrical signals around the body so your brain can communicate with all the other organs.
3 The main cell body of a nerve cell is spiky and has connections called *dendrites*. These allow one nerve cell to be connected with many others.
4 Each nerve cell is very long – some can be up to a metre in length. They have a long tube called an *axon* that ends with nerve endings. Signals start at the dendrites and get sent all the way to the nerve endings in a fraction of a second.

Check for understanding

Display the *CFU* slide and, using MWBs, ask questions like the ones below.

- What is the function of the nerve cell?
- Give one adaptation of the nerve cell.
- Does a nerve cell contain a nucleus/mitochondria/cell membrane/cytoplasm?
- Why are nerve cells very long?
- Why do nerve cells have many dendrites?

Independent practice review

Student Practice Book questions 59–75 (page 35).

Q	Notes
63, 64	These questions are aimed to help students make links between previous topics. Students might need a verbal hint to remind them to think back.

Guided explanation

What are tissues?

Draw the diagram below sequentially, with each step accompanied by an explanation, similar to the one that follows.

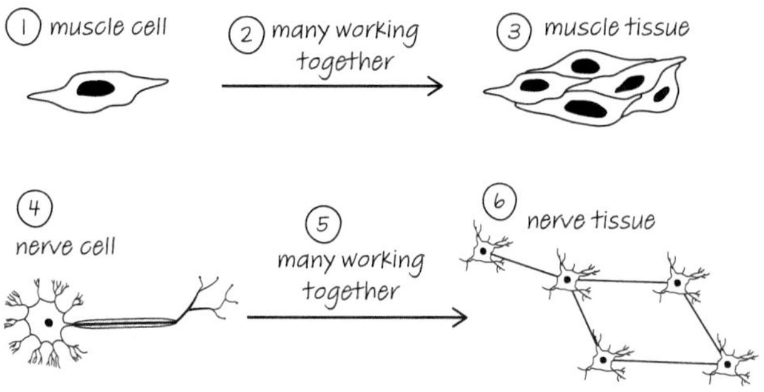

1 Tissues are groups of the same cell working together. For example, if we have a muscle cell we know it is specialised to contract to allow movement, but one cell is not very strong.
2 If we get lots of muscle cells and connect them together then they can pull with much more force. Tissues allow cells to increase their effect.
3 We call this muscle tissue.
4 It is the same with all specialised cells – e.g. our nerve cells.
5 When we connect lots of them together they work better.
6 They make nerve tissue, which allows us to send signals all over the body.

Check for understanding

Display the *CFU* slide and, using MWBs, ask questions like the ones below.

- What do we call a group of specialised cells working together?
- Why do muscle cells form muscle tissue?
- Why do nerve cells form nerve tissue?
- Why do specialised cells form tissues?

Independent practice review

Student Practice Book questions 76–85 (page 36).

Q	Notes
76	Students need to be clear that, by combining, the cells can do more work or apply a stronger force.

Guided explanation

What is an organ?

Explain that the heart is one of the most important organs in our body. The heart's function is to pump blood around the body, to ensure all cells have the oxygen and nutrients they need to survive.

Display the slide and add to the diagram, as shown. Construct the diagram sequentially, with each step accompanied by an explanation, similar to the one that follows.

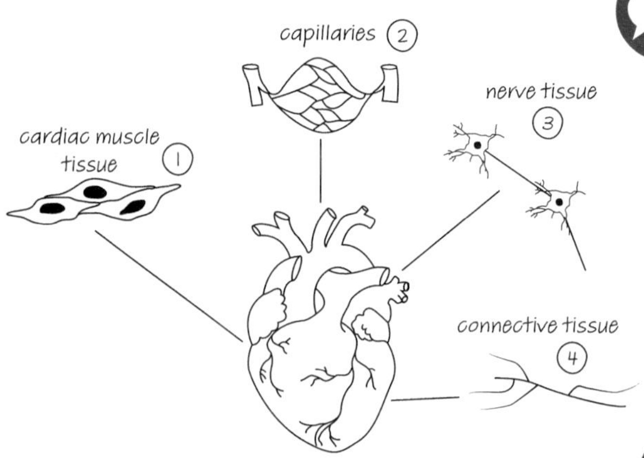

1 The heart is an organ because it is made of more than one tissue. The heart is a muscle, so it has muscle tissue in it. The cardiac muscle cells form cardiac muscle tissue to contract and pump the blood.
2 The muscle tissue needs its own supply of nutrients and oxygen, so in the heart there are blood vessels called *capillaries*, which are a tissue. They ensure the supply of blood to the cardiac muscle tissue.
3 Sometimes your heart needs to beat faster and sometimes it needs to beat slower. The heartbeat is controlled by the brain. Therefore, the heart also has nerve tissue running through it to control the speed at which the heart beats.
4 For all these tissues to function they need to be in the right place. Your heart has connective tissue, which holds everything together in a strong but flexible form.

Say that, in general, organs are groups of tissues working together to perform the same function. In this case, the heart is an organ because the tissues are all working together to pump the blood.

Check for understanding

Display the *CFU* slide and, using MWBs, ask questions like the ones below.

- Name two tissues in the heart.
- Which tissue is responsible for pumping the blood?
- Which tissue is responsible for ensuring the supply of nutrients?
- Which tissue controls the rate of heartbeats?
- Which tissue holds all the other tissues together?
- Why is cardiac muscle called a tissue?

Independent practice review

Student Practice Book questions 86–101 (pages 36–37).

Q	Notes
89	Make sure they have the correct reasons for each adaptation.

Guided explanation

How are structures within an organism organised?

Tell your students that we will now put everything together that we have learned so far. Draw the diagram below sequentially, with each step accompanied by an explanation, similar to the one that follows.

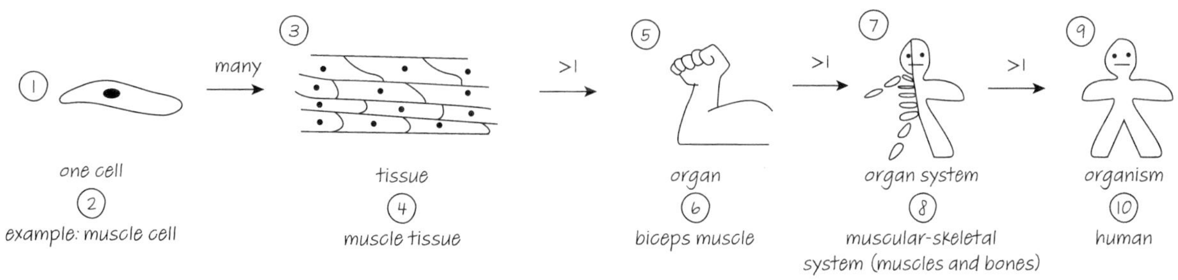

1 We start with a single cell, the smallest unit of life. This cell is specialised in a multicellular organism.
2 For example, a muscle cell.
3 If many specialised cells join together, they form a tissue.
4 For example, muscle tissue.
5 When more than one tissue joins together, they form an organ.
6 In this example, the muscle tissue and the connective tissue work together to form a muscle, like the biceps.
7 All the organs that help to do the same function form an organ system.
8 In this example, all the muscles and bones form together in the muscular-skeletal system.
9 All the different organ systems, like respiratory, reproductive and digestive, work in harmony to form an organism.
10 An organism, like you, is a group of organ systems.

Check for understanding

Display the *CFU* slide and, using MWBs, ask questions like the ones below.

- What is a group of specialised cells called?
- What is more than one tissue working together called?
- What is more than one organ working together called?
- What do we call all the organ systems working together?
- The heart is an organ, because it is made of more than one *what*?
- What layer of organisation does A show?
- Which letter is the organ?
- Which letters correspond to layers that need more than one type of cell?

Independent practice review

Student Practice Book questions 102–11 (pages 37–39).

Q	Notes
104	Students often confuse an organ with an organ system. This question is designed to help remove that problem. However, this might need regular checking.

C1 The particle model

C1.1 Simple particle model

Students' learning objective: *I am learning about the simple particle model of matter so I can explain the properties of solids, liquids and gases.*

This is the first chemistry unit and builds on student understanding from KS2, where they will have investigated states of matter and classified objects based on their state.

Knowledge check

As this is the first chemistry topic, there is no *General science quiz* at the start. Display the *Knowledge check* slide. The questions are best delivered one by one on mini-whiteboards (MWBs). If students do not know the answer to any of them, you should reteach the relevant concept. Answers are not on the slides because it is best to go over them together as a class.

	Question	Answer	Notes
1	What happens to water when put in a freezer?	It gets colder or it freezes	These two questions are to check that students know that ice and water are the same substance in different states.
2	What do we call water in the solid form?	Ice	
3	What happens to water when it is heated in a kettle or on a stove?	It gets hotter or it boils	These two questions are to check that students know that water and steam are the same substance in different states.
4	What do we call water in the gas form?	Steam	
5	What is the difference between the three forms of water?	Temperature	This question is to check that students understand that the only difference between ice, water and steam is the temperature. Often students think these substances are chemically different or are made of different substances. It is therefore important to check for, and challenge, this potential misunderstanding in all students.

Guided explanation

What is the particle model of matter?

Nobel-prize-winning physicist Richard Feynman (particle physicist and part of the Manhattan project) was once asked 'What single sentence would be able to kick-start science if all scientific knowledge were lost?' The following was part of his response: 'All things are made of little particles that move around in perpetual motion.'

Display the first slide. Do not ask students to guess the origin of this sentence or what it means, but do check that they understand the meaning of the words *perpetual* and *motion*, which may need defining with some students.

Feynman's response would be a useful statement to restart science because it is important for understanding many of the ideas that are coming later, which is why we are learning it in the first topic. At the same time, it is something that we cannot see easily with our own eyes, or even with a microscope, because the particles that make up everything are so small.

The particle model

Tell students that all physical objects are made from particles, and give them three or four examples, including gases, the chairs they are sat on, the air they are breathing, the shirts they are wearing.

Using water as an example, introduce the states of matter by drawing the diagram below, accompanied by an explanation, similar to the one that follows. It may be useful to have examples of water, ice and a kettle to make steam available as a demonstration.

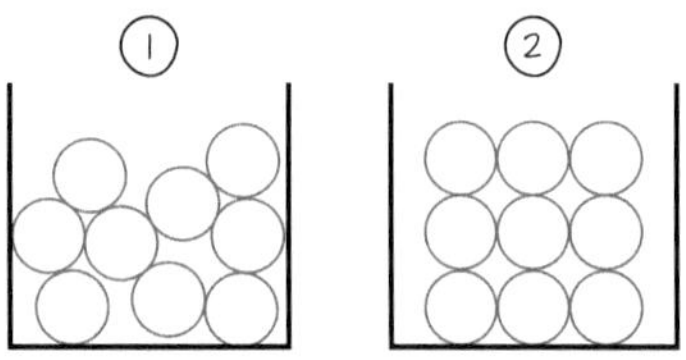

1 Draw the first diagram and tell students that it shows the particles in a glass of water.

2 Add the second diagram to show the particles in an ice cube. Reinforce the idea that the particles are the same (so draw them the same size), but they are just arranged differently.

Move from the specific example of water to a general example of how to draw particle diagrams for solids, liquids and gases. Students are likely to need the word *regular* defined for them in this context.

Draw the diagram below to show the arrangement of particles in all three states. As you construct the diagram, accompany each state with an explanation, similar to the one that follows.

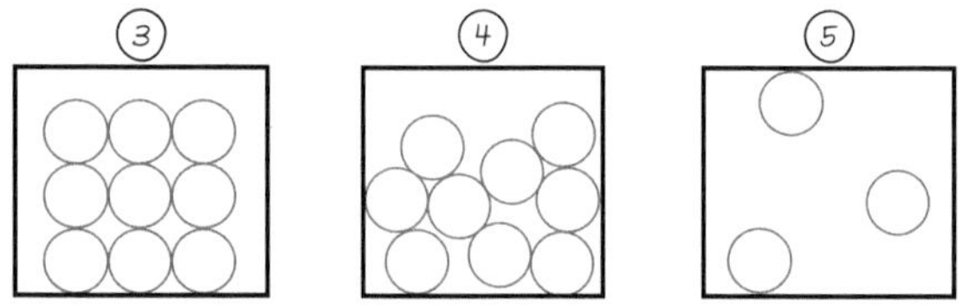

3 The arrangement of particles is caused by the forces acting between the particles. In solids the particles have strong forces holding them together, so that the particles are only able to vibrate – they are not able to break free from their neighbouring particles.

4 In liquids the particles are able to break free from some of the attractive forces with their neighbours, so that they can slide past each other, but they cannot break free of all the forces of attraction, which is why they are all still touching each other.

5 In gases, the particles have completely broken free of the attractive forces with each other, so each particle moves independently and without touching any others; unless they bump into each other – which they do, a lot! The particles in a gas therefore have the weakest forces of attraction between the particles.

Check for understanding

The Check for understanding (CFU) is best conducted using MWBs. Ask questions like the ones below. Tailor and adapt your questions to the class in front of you.

When asking students to draw particle diagrams, ask them to draw a box in the centre of their board first and to draw their diagram in that box. When moving on to the next diagram, tell them not to rub out the box, e.g. 'Without rubbing out the box, clean your board. On your MWB, draw a particle diagram for a gas.'

- Draw a particle diagram for a solid. *Do not accept different-sized particles. Look out for diagrams that are in the middle of the box instead of at the bottom.*
- Draw a particle diagram for a liquid. *Do not accept different-sized particles. Look out for diagrams that are in the middle of the box instead of at the bottom. Challenge large gaps between the particles that have room for more particles.*
- Draw a particle diagram for a gas. *Do not accept different sized particles. Challenge 'movement lines', which are not desirable when drawing the arrangement of particles.*
- Display the *CFU* slide and draw a poor particle diagram. Ask what is wrong with the diagram. *This question can be repeated to pick up on errors you noticed in the earlier CFU questions.*

Independent practice review

Students practise independently by answering questions 1–7 from their Practice Books (page 40). Full answers can be found on the *IP answers* slides or in the answer documents on Boost. Notes are only included for questions, where useful.

Q	Notes
2	It is important that students are secure in the knowledge that the particles are unchanged between states, and the only differences are in the arrangement and movement of particles.
4	These non-examples are a useful way for students to challenge their own misconceptions and potential errors in their own diagrams.
6	This question checks that students understand that the arrangement of particles is caused by attractive forces between particles, and these are strongest in solids since the particles cannot move past each other and can only vibrate.
7	A common misconception in students is that particles in a solid do not move at all. This refutation question challenges students to explain why this is incorrect.

C1.2 Properties of different states of matter

Students' learning objective: *I am learning about the properties of states of matter so I can explain their behaviour.*

Students may be familiar with properties of objects from KS2, although this focus is more likely to be on bulk, physical properties of objects, rather than the properties of specific states of matter.

Knowledge check

After working through the *General science quiz*, display the *Knowledge check* slide and, using MWBs, deliver the questions one by one, reteaching concepts as necessary.

	Question	Answer	Notes
1	What are the three states of matter?	Solid, liquid, gas	Accept any order.
2	Draw a particle diagram for a solid.	Appropriate diagram meeting the following criteria: ● contains at least nine particles ● regular arrangement ● touching ● at the bottom of the box ● no overlaps ● all particles roughly the same size	Look out for: ● particles not touching ● particles not in a regular pattern ● particles of different sizes
3	Describe the arrangement of particles in a liquid.	Random arrangement, touching	Do not accept alternative words for 'random' (e.g. muddled, higgledy-piggledy). Do not accept students giving the key features of the diagram instead of the arrangement of the particles.
4	Describe the arrangement of the particles in a gas.	Random, not touching	Do not accept alternative words for 'random'. Do not accept students giving the key features of the diagram instead of the arrangement of the particles.
5	How do the particles in a gas move?	Randomly, in all directions	Do not accept fast-moving or other synonyms for quick. Some particles move fast, but the speed of particles is also random. At this stage focus attention on the random directions of movement.
6	How do the particles in a solid move?	They vibrate around a fixed point	Accept other correct phrasings, but do not accept 'wobble'.

Guided explanation

Which substances can flow and why?

There is a lot of abstract information in this part of the learning sequence, so the explanation is broken into several sections, with a Check for understanding and Independent practice review between each one.

What was Democritus' particle idea?

Introduce the topic by giving students some historical background information. Display the first slide and describe the particle theory of the Greek philosopher Democritus (though not his 'atomic theory', which would confuse things now while we are just dealing with 'particles').

Democritus was a Greek philosopher who noticed, while walking along the beach, that the rocks, pebbles and sand were all made of the same material, but simply in smaller and smaller pieces. He also noticed that as the pieces of solid material became smaller in the sand, they behaved slightly differently from the rocks, and behaved a little like the water in the sea. He noticed that if the sand was dry, then it took on the shape of its container and could flow through his fingers, like the water in the sea. Democritus thought that perhaps the water in the sea was also made of small particles, like the sand, but that unlike the sand, the particles of water were too small to be able to see.

1 Display the second slide with pictures of sand and water flowing through people's hands. Ask students to use their own personal experience of sand and water, and to think like Democritus did.

Ask them in which ways sand and water are similar, and in which ways they are different.

2 Put students in pairs for a discussion activity. This activity works well when students have 30 seconds of silent time to think for themselves, then some discussion time with their partner. Ask the class to share their ideas by asking discussion partners to discuss each other's ideas. Student ideas to highlight are:
   - those about properties (flowing, keeping shape etc.)
   - those about particle size
   - each grain of sand is actually a solid, made of many particles.

3 Alternatively, you could use *Front the Writing*, with everyone jotting their ideas down for 90 seconds in silence first, before discussing them with their partners. This helps to build rationale through writing.

4 Encourage very good student responses, which identify that the key difference is that sand particles are *not* moving, which is why sand does not behave as a liquid. Fine sand, however, will behave like a liquid if its grains are constantly moved by an external force (vibration or air movement).

Which substances can flow and why?

Draw the particle diagram below sequentially, with each step accompanied by an explanation, similar to the one that follows.

1 Describe how, if you put a solid (e.g. ice) into a cup, it would stay ice-cube shaped – it would not take the shape of the container, whereas the liquid would. You may want to draw a diagram showing this, or have an ice cube to demonstrate the phenomenon.
Explain that the reason the solid keeps its shape, and has a regular arrangement, is because the particles are held very tightly together – it has strong forces of attraction between the particles. For some groups, it might be useful to write this phrase underneath the solid particle diagram.

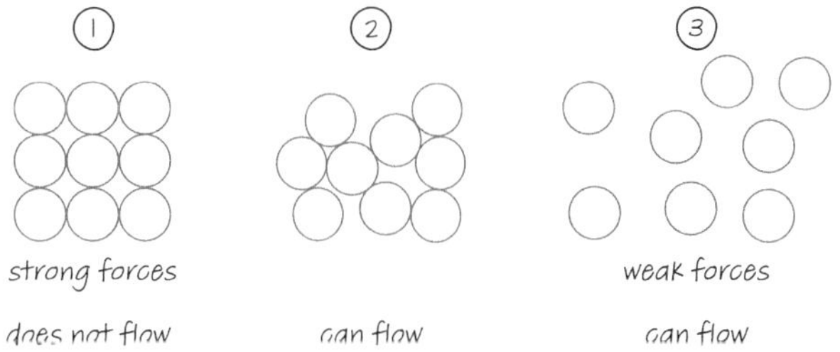

2 Focus now on the liquid diagram, reminding students that you can pour a liquid. Explain that this is because the particles can move over and past each other, and that this is called *flowing*.
Explain that the reason liquids can flow and have this random arrangement is because they have weaker/medium-strength forces of attraction between the particles.

3 Add the particle diagram for a gas. Make sure to use the same number of particles as you did for the solid and the liquid.
Explain that the particles are the same particles; they are just further apart, and the reason they are further apart is because they have much weaker forces of attraction between the particles. Again, for some groups, you may want to write this under the diagram.

Check for understanding

Display the *CFU* slide and, using MWBs, ask questions like the ones below. Avoid asking explicitly about *properties*, as this is not a word that students have been taught to define yet.

- What is flowing?
- Which state or states can flow? *Accept either order. If students write 'fluids', prompt them further to check they know what fluids are.*
- Why can these states flow?
- Why can't solids flow?

Independent practice review

Student Practice Book questions 8–10 (page 40).

Q	Notes
10	This refutation question challenges a common misconception. Students need to understand that each rice gain is a solid consisting of billions of particles, and any individual grain behaves entirely as a solid. As a collection of small solids, grains of rice – like sand – can be poured and flow through your fingers, but they form a pile rather than completely taking the shape of their container, this is because each grain is not moving and flowing over the others as particles do in a liquid.

Guided explanation

Which substances can be compressed?

Draw the particle diagram of a gas below sequentially, with each step accompanied by an explanation, similar to the one that follows.

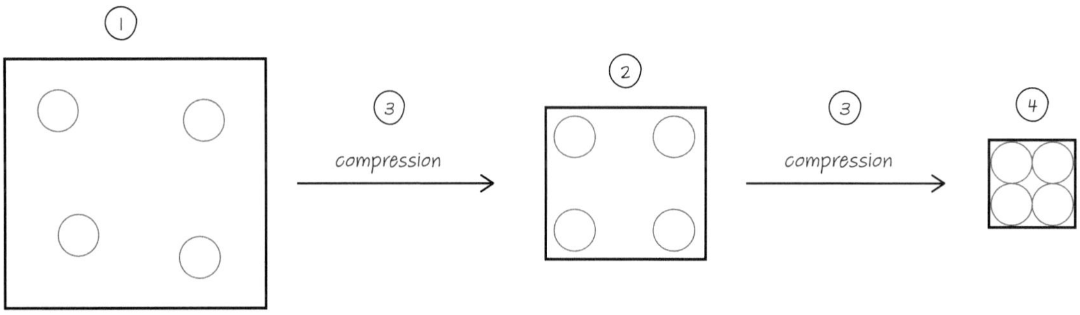

1 Point out to students that there is space between the particles.
2 Add the smaller box. Explain that you have made the box smaller. You can say that you have 'squeezed' or 'squashed' the box. Point out that there is still space between the particles, but that this space is now much smaller.
3 Explain that this 'squeezing' or 'squashing' is called *compressing*, and that you have *compressed* the gas. You may want to draw an arrow between the boxes, labelled *compression*, as shown. Make sure that students are clear that the particles themselves have not been changed.
4 Draw the particle diagram for a solid. Explain that there are no spaces between the particles, so we cannot compress solids.

For some groups, you could do this section using a short series of questions, such as those below, asking students to put up their hands to answer.

- Are there any spaces between the particles? *No*
- So will we be able to compress it? *No*

Check for understanding

Display the *CFU* slide and, using MWBs, ask questions like the ones below.

- Define *compression or* What is compression?
- Why can gases be compressed? *Ignore references to forces.*
- Why can't solids/liquids be compressed?

Independent practice review

Student Practice Book questions 11–12 (page 41).

Q	Notes
11	Students should be secure in their knowledge that solids and liquids have no significant gaps between particles.
12	*Because-But-So* writing stems are a good way of elucidating students' understanding and their ability to link concepts and ideas. There is not necessarily one 'right' answer; indeed, there may be many correct answers. There can of course be incorrect answers, and therefore during circulation it may be a good idea to highlight mistakes as well as identifying student work for *Show Call*, showing, discussing and marking some responses under the visualiser for the whole class to see. ● Because – indicates why something is true. ● But – indicates a change of direction; perhaps not true in all cases. ● So – indicates a consequence of this; a cause and effect. **a** The particles in a liquid are free to move over one another because ... *[particles have more energy than when a solid/particles are less strongly attracted to each other].* **b** The particles in a liquid are free to move over one another, but ... *[the particles are still all touching each other].* **c** The particles in a liquid are free to move over one another, so ... *[they flow and take the shape of their container].* **d** The particles in a gas are free to move randomly in all directions because ... *[the particles have lots of energy/the particles move fast/the particles have weak forces of attraction between each other].* **e** The particles in a gas are free to move randomly in all directions, but ... *[they bump into their containers and each other/they move in straight lines until they bump into another particle or their container].* **f** The particles in a gas are free to move randomly in all directions, so ... *[they fill their container/they flow].*

Guided explanation

What are properties?

Highlight to students that they have looked at whether something can flow and whether it can be compressed. These are things something can do, or how it behaves.

1 Explain that these are called *properties*.

2 Give students some examples of properties that they may be familiar with, such as whether something can conduct electricity, or if it is hard, or what colour it is. Explain that these are normally things that we can see or easily find out through our senses. Write these as a list on the board.

3 Tell students that the arrangement of particles is not a property – we cannot see the particles in an ice cube being arranged all regularly. Write *arrangement of particles* in a separate list. Add the heading *properties* above the list of properties. Explain that the arrangement of particles is the *structure* and write that as a heading.

4 Tell students that *strong forces of attraction*, or equivalent, is also not a property, and that it is something called *bonding*. Write this in a third list and add the heading *bonding*, to end up with the following on the board:

Properties	Structure	Bonding
conducts electricity	arrangement of particles	strong forces of attraction
hard		
colour		

5 Explain that scientists can use the bonding to explain the structure, and then the properties of something. Give students an example.

A summary of this part of the explanation can be found in Topic C1.2 of the Knowledge Book.

Check for understanding

Display the *CFU* slide and, using MWBs, ask questions like the ones below.

- What are properties? *Ensure students are not giving descriptions of the particle arrangements or their temperature etc.*
- Give two properties of gases/liquids/solids. *Do not accept references to bonding or structure.*
- Why are terms such as 'has strong bonds' or 'regular arrangement' not properties?

Independent practice review

Student Practice Book questions 13–17 (page 41).

Q	Notes
13	This refutation question challenges a common misconception. Students need to understand what 'properties' are, and that they are not descriptions of the movement or arrangement of particles. The corrected text could either correct the words 'one property' to 'particles in solids' or change the end of the sentence to a property of solids, such as 'cannot be compressed'.
16	This refutation question challenges a common misconception that particles can change size and that the particles get smaller when something is compressed.
17	This is a challenging question, but students who grasp this concept well will be able to realise that once all particles are touching the substance is, by definition, no longer a gas but is in liquid form. This is what happens in compressed gas cylinders.

C1.3 Changes of state

Students' learning objective: *I am learning about changes of state so I can describe what happens to materials when they are heated or cooled.*

Students will be familiar with changes of state from KS2. However, they may not be as familiar with all four state changes, or with the relationship between energy and change of state.

Knowledge check

After working through the *General science quiz*, display the *Knowledge check* slide and, using MWBs, deliver the questions one by one, reteaching concepts as necessary.

	Question	Answer	Notes
1	Draw the arrangement of particles in solids, liquids, and gases.	As in previous topics	As in previous topics
2	Describe the movement of particles in a solid.	Particles vibrate around a fixed point	Accept answers that are not in full sentences.
3	Describe the movement of particles in liquids.	Particles slide over each other, but remain touching	Both parts are required for a correct answer.

	Question	Answer	Notes
4	Describe the movement of particles in gases.	Particles move randomly in all directions	
5	What is the difference between water particles in ice and water particles in steam?	The particles are the same – the difference is in their arrangement and their movement	If students have only given the first part, prompt them to expand to include the second part.

Guided explanation

What are the changes of state?

Start with the familiar example of water. Draw the diagram shown below sequentially, with each step accompanied by student input or explanation, similar to the outline that follows.

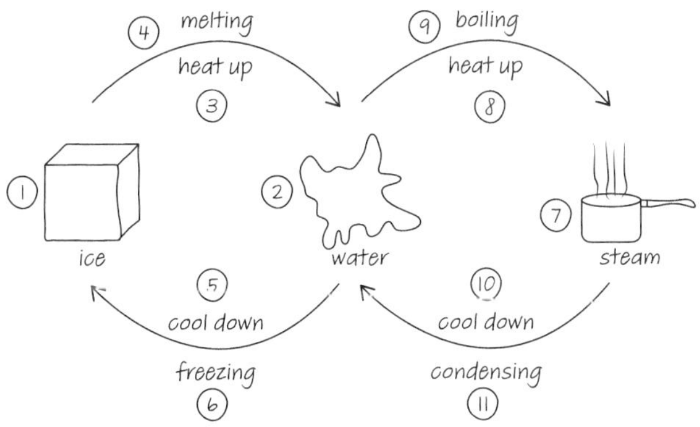

1 Draw and label an ice cube.
2 Draw and label a puddle of water.
3 Tell students that to get from ice to water we need to heat up the ice cube. Add a labelled arrow pointing from the ice cube to the puddle.
4 Ask if students know what this process is called. Add the *melting* label above the arrow you have just drawn.
5 Tell students that to get from water to ice we need to cool it down. You can phrase it as 'cool down', or with some groups you might prefer to centre it as 'heat out', which makes the transition to understanding internal energy at KS4 easier. Label the arrow with your choice of phrase.
6 Ask if students know what this process is called. Add the *freezing* label below the arrow you have just drawn.
7 Draw and label a steaming saucepan. Add an arrow from the puddle to the steam.
8 Ask students how you get from water to steam. Add the labelled arrow pointing from the puddle to the saucepan.
9 Add the label for *boiling* above the arrow you have just drawn. Do not ask students to name this process, because some will give you 'evaporation', and it is important to avoid confusion between these two processes. If a student mentions evaporation, tell them it is a bit of a special case, and you will come back to it.
10 Draw an arrow from the steam to the water. Tell students that we have to cool down steam to get water.
11 Say that this process is called *condensing* and add a label for this.

Explain that these processes happen with other substances, not just with water. Erase the word 'ice' and replace it with 'solid', erase 'water' and replace it with the word 'liquid', and finally remove the word 'steam' and replace it with 'gas' – to finish with the diagram shown below.

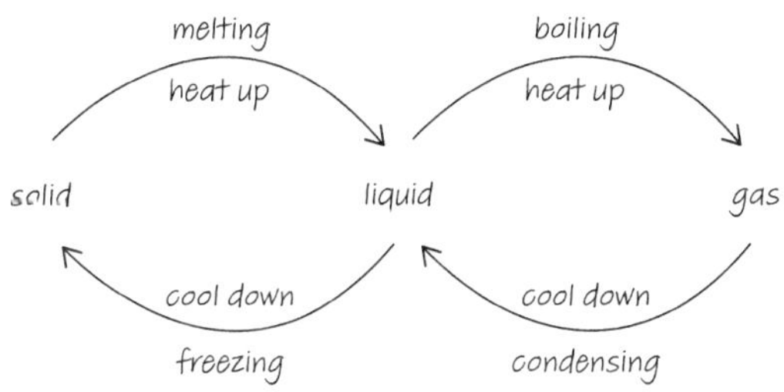

Check for understanding

Display the *CFU* slide and, using MWBs, ask questions like the ones below.

- What is A? (*Repeat for B and C.*)
- What process is happening at D? (*Repeat for E–G.*)

There is potentially a lot of new information presented to students in this topic, and with some groups, you might find it useful to leave the diagram complete and visible, asking questions such as: *What is it called when a solid turns into a liquid?* After students have sufficiently consolidated their knowledge, you can then remove the words and continue as above to truly check for understanding.

Independent practice review

Student Practice Book questions 18–24 (page 42).

Q	Notes
18–20	As well as checking understanding of the terms 'melting' and 'boiling' and the arrangement of particles, students should be encouraged to check that their particles are the same size in all three diagrams, since these questions are all about the same substance changing state.
23, 24	As liquids are heated the small spaces between particles increase, slightly, but students should be reminded that there are still no gaps large enough for extra particles, and the substance cannot be compressed.

Guided explanation

What are melting and boiling points?

Explain that melting and boiling happen at specific temperatures for specific substances. Give students a few examples, starting with familiar ones such as water, but include examples that are solids at room temperature and ones that are gases at room temperature. Some suggestions are given below, along with their melting and boiling points.

Substance	Melting point (°C)	Boiling point (°C)
Wax (paraffin)	37 (approx.)	370 (approx.)
Iron	1538	2862
Oxygen	-218.8	-183

Make sure you use the word 'pure' in your examples. If a student asks what pure means, do not mention atoms – they have not learned this yet. Instead, say 'it's when you have just one thing and nothing else in it, so just water, for example'.

Draw a vertical line labelled *solid*, *liquid* and *gas*, as shown below. Then add the other labels sequentially, with each step accompanied by an explanation, similar to the one that follows.

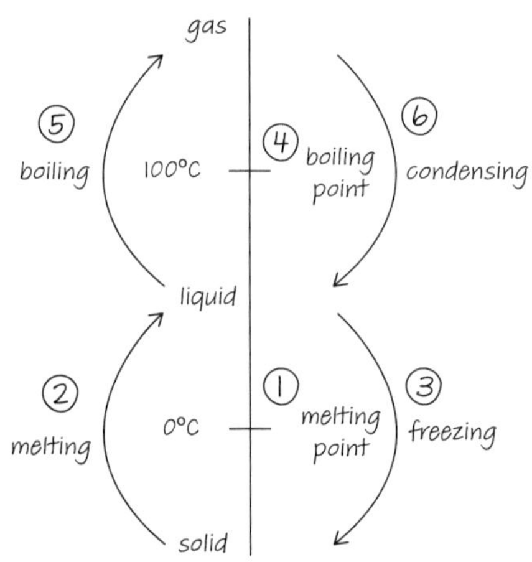

1 Draw and label a short horizontal line for *melting point*.
2 Add an arrow labelled *melting* that goes from *solid* to *liquid*.
3 Draw an arrow labelled *freezing* opposite this, from *liquid* to *solid*. Explain that substances melt at the same temperature that they freeze at, and that this is called the *melting point*. The temperature remains the same until the substance has melted.
4–6 Repeat this for boiling point, boiling and condensing.

Tell students that you can use the *melting* and *boiling points* to work out what state something will be at a given temperature. Show this by adding the values 0°C and 100°C for water. Then erase these and, one at a time, add values from the earlier examples, showing where room temperature is on each number line to illustrate why these substances are different states when we encounter them normally.

Check for understanding

Display the *CFU* slide and, using MWBs, ask questions like the ones below.

- At what temperature does water freeze? *Do not accept '−1°C' or 'less than 0°C'. Do not accept answers that are lacking the complete unit (e.g. '0°' or '0').*
- At what temperature does steam condense? *Do not accept '99°C' or 'less than 100°C'. Do not accept answers that are lacking the complete unit (e.g. '100°' or '100').*
- Oxygen has a melting point of X and a boiling point of Y. (*Write these values on the board, so students do not have to keep them in their head at the same time as working out the answer.*) What state would it be at X°C? What about at Y°C?

The last question is a tricky concept for students to understand, and you may need to repeat this task several times with several substances to be confident that students understand. It is useful to model how students should approach these problems:

- Draw a number line.
- Add the melting point and boiling point.
- Direct attention to the bottom of the number line (including negative numbers), rising up it until the given temperature is reached and identifying whether the melting points or boiling points have been passed.

Independent practice review

Student Practice Book questions 25–31 (pages 42–43).

Q	Notes
25	Check that students understand that the temperature does not change during a state change.
27	A common misconception is that once a substance becomes a gas its temperature no longer rises further. Check that students understand that gases can increase further in temperature.
28	This question checks that students understand that melting–freezing and boiling–condensing happen at the same temperature; the only difference is whether the substance is being heated or cooled.
31	Some students will find the use of negative numbers challenging, and you should be looking out for students who need support with a number line featuring negative numbers.

Core practical

Heating water

A method is provided on the *Practical worksheet* and *Teacher and technician notes* on Boost – including equipment list and safety notes – if you wish to allow your students to proceed independently. However, for most students it will ordinarily be better to take a *Slow Practical* approach following the guidance below.

Students should ideally work in groups of two, but no more than three. For each group and yourself, you will need eye protection, a Bunsen burner, heatproof mat, tripod, gauze, 250 cm^3 beaker, thermometer, stopwatch and clamp with stand.

Set up the demo station at the front with the heating equipment as shown on the first slide, but do not light the Bunsen.

Display the second slide, showing a blank template for the hazards and precautions table below.

② Hazards	③ Precautions
① burns from Bunsen	safety flame
burns from water	don't turn your back
broken glass	away from edge of bench
	don't move when hot

1 Ask students to tell you how someone could get hurt when heating something with the equipment shown.

2 If their suggestions are correct, write them into the 'hazards' column of the table. Add any further hazards involved with heating. Tell students that we call these things that could harm them 'hazards'. Write this as the column heading.

3 To stay safe, we need to make sure we do certain things. We call these things 'precautions'. Add this as the second column heading. For each of the identified hazards, ask students what they could do to keep themselves and others safe. Add to their suggestions, as needed, to ensure a safe practical. (Hazards and risks will be covered in more detail in Unit C2.)

4 Fill a 250 cm^3 beaker with approximately 100 cm^3 water. It is not necessary to measure this precisely.

5 Assign the equipment by getting students to collect apparatus as needed and return to their workstations. Once students have set up their equipment correctly, light each group's Bunsen.

6 Students should measure the temperature every minute and record their results in the table on the *Practical worksheet* until the water reaches approximately 100°C.

7 They continue to heat the water for a further 5 minutes, recording their results. You will find that some thermometers do not read 100°C. You may want to talk about error, but this is not necessary – you can simply say the thermometer is not accurate.

8 Highlight to students that once the water reached its boiling point, the temperature no longer increased.

9 Ask students to complete the independent practice questions on the *Practical worksheet*.

Guided explanation

What is the difference between evaporation and boiling?

Draw a T-shirt with a particle diagram of a liquid on it. Construct it sequentially, with each step accompanied by an explanation, similar to the one that follows.

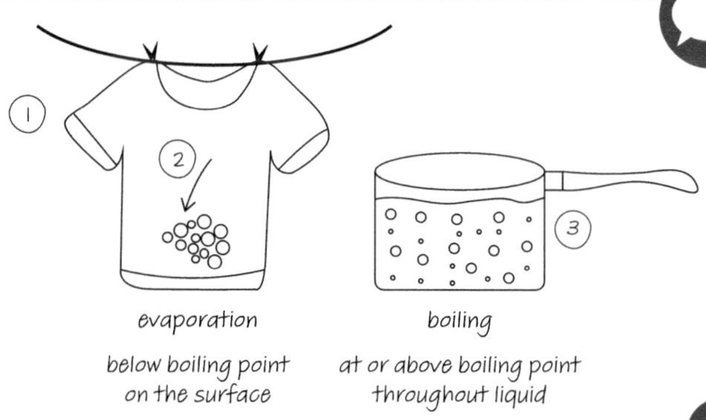

1 Explain that when their clothes dry, the water is not boiling off. It is not hot enough for the water to boil.

2 Draw an arrow pointing at one of the 'top' particles. Explain that sometimes an individual particle will have been heated up enough to be able to move away from the rest of the liquid, and we say it has *evaporated*. It is more accurate to speak in terms of energy, but at this point students might not have learned about energy.

3 Draw a saucepan full of water. Draw bubbles throughout the liquid. Make these are different sizes, and ideally a different colour to make it clear that these are not particles. Explain that boiling and evaporation are similar – they are both a liquid turning to gas – but that boiling happens throughout the whole liquid and only happens at or above the boiling point, whereas evaporation can happen below the boiling point, and only happens on the surface.

Check for understanding

Display the *CFU* slide and, using MWBs, ask questions like the ones below.

- How are boiling and evaporation different? *Reference to boiling and evaporation are needed.*
- How are boiling and evaporation similar? *Do not accept 'water to steam', or other specific examples.*
- Where does evaporation happen? *Look out for spatial answers, such as 'outside' or 'when it is sunny', which may arise because of the everyday examples given so far.*
- Does a substance have to be at the boiling point for evaporation to take place? Why/why not?

Independent practice review

Student Practice Book questions 32–34 (pages 43–44).

Q	Notes
32	The key point here is that the water particles still exist, in the form of steam. Challenge any misconceptions that the water particles have been 'burned' or destroyed in any way. Students will likely consider that the water has boiled, but the water will have turned into a gas by both evaporation and boiling.
34	This refutation question challenges the misconception that the only way that a liquid turns into a gas is by boiling.

C1.4 Gas pressure

Students' learning objective: *I am learning about gas pressure so I can predict how different conditions affect gas pressure.*

This topic is not one students will have any familiarity of from KS2.

Knowledge check

After working through the *General science quiz*, display the *Knowledge check* slide and, using MWBs, deliver the questions one by one, reteaching concepts as necessary.

	Question	Answer	Notes
1	In which state (or states) of matter can particles move freely?	Gases	Do not accept gases and liquids, because whilst the particles in liquids can move over one another, they cannot move freely – i.e. away from their neighbour.
2	In which state (or states) of matter are particles touching?	Solids and liquids	Accept either order.
3	What happens to particles when you heat them up?	They move faster	Do not accept 'change state' – it is the whole substance that changes state, not individual particles
4a	Which state (or states) of matter fill their containers?	Gas	Liquids take the shape of their container, but do not fill it.
4b	Explain your answer to the previous question.	Particles in gases move very fast and have overcome the forces holding them close to their neighbour, so can travel everywhere in their container	

Guided explanation

What is gas pressure?

Draw a diagram of a foil balloon with gas particles inside, as shown below. As you construct the diagram, accompany it with an explanation, similar to the one that follows.

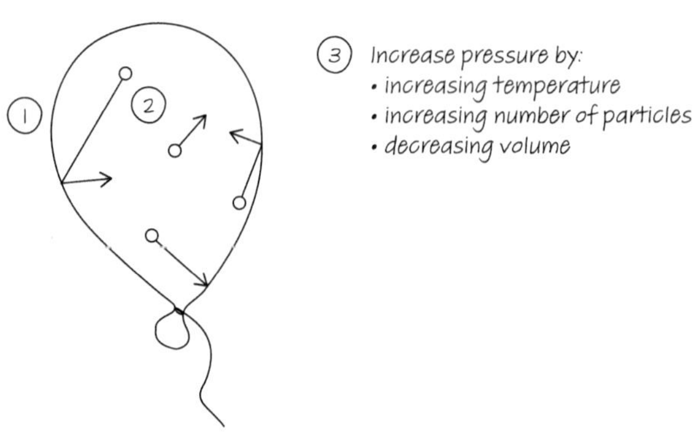

1 Remind students that the particles in a gas move randomly and in all directions.

2 Add some arrows to the particles to demonstrate motion. Tell students that this means they will eventually hit the surface of the container. Continue adding arrows that collide with the balloon. Tell students that we say these particles have *collided* with the surface of the container, and when they hit (collide with) the container, they push against it a little bit. We call this pushing from these collisions *pressure*. At this point students have not learned about forces, so using the language of 'pushing' allows students to understand the concept, without being confused about what a force actually is. Forces are described as a push or a pull, so using the word push does not risk embedding misconceptions.

3 Explain that we can change the pressure. Remind students that pressure is caused by particles colliding with the surface of a container, so if we want to increase the pressure we need to increase the number of collisions, or how hard those collisions are. Tell students that there are three ways of doing this – heat it up, add more particles or make the space smaller.

For some groups, you will be able to take suggestions for why each of those things increases the pressure, but for other groups, you will need to explain how each of those things increases the number of collisions:

- Explain that if we heat up the particles they move faster. Then explain that this means they will collide with the surface more frequently (there is no need to get into discussions around energy at this point).
- Explain that adding more particles means there will be more collisions. You may want to add more particles to your diagram to illustrate this.
- Explain that making the volume smaller means the particles will collide with the sides more frequently as well. You may want to draw a smaller foil balloon with the same number of particles to illustrate this.

Check for understanding

Display the *CFU* slide and, using MWBs, ask questions like the ones below.

- What are the three ways to increase the pressure? *Accept any order.*
- What will happen if ... e.g. I reduce the number of particles/cool the space down/make the space larger? *Be careful to ask questions about how to both increase and decrease the pressure.*

Independent practice review

Student Practice Book questions 35–44 (page 44).

Q	Notes
37, 38	This is a key question for students to demonstrate understanding because it links the ideas of speed of particle movement changing with temperature and pressure. It is also phrased in reverse compared with explanations that are in the context of increasing pressure, with this relating to a decrease in pressure.
41	See page 55 for an explanation of how to implement *Because-But-So* activities. **a** If a gas is heated up, the particles ... [move faster] because ... [they have more energy]. **b** If a gas is heated up, the particles ... [move faster] but ... [do not change size or mass]. **c** If a gas is heated up, the particles ... [move faster] so ... [the pressure increases]. **d** If a gas is heated up, the pressure ... [increases] because ... [the number of collisions with the container increases]. **e** If a gas is heated up, the pressure ... [increases] but ... [the number of particles inside the container is the same]. **f** If a gas is heated up, the pressure ... [increases] so ... [the container may break if the pressure becomes too high]. **g** If a gas is cooled down ... [the pressure decreases because the particles move slower and have fewer collisions with their container].
43–44	These are real-world applications of gas pressure and a useful opportunity to check that students understand the dangers of pressurised gas containers.

C2 Atoms, elements and compounds

The National Curriculum point *conservation of mass, changes of state and chemical reactions* is not covered in this unit. The content is covered in Unit C4, because students are required to know the nature of chemical reactions before they can understand the conservation of mass in chemical reactions and state changes.

C2.1 The atomic model

Students' learning objective: *I am learning about the different types of atom so I can explain the layout of the Periodic Table.*

The aim of this topic is to learn the fundamental basis of atoms and atomic theory and the language of chemistry and chemical reactions. Students will also learn the modern model of the atom, and how this is related to the structure of the Periodic Table.

Students may have a wide range of prior knowledge and misconceptions. Some of them may have heard some of the terms used in this topic before, but may not be using them correctly. Most students will have seen a Periodic Table, but will not be totally sure about what it contains or how it is arranged.

The National Curriculum refers to a *simple (Dalton) atomic* model. Within this topic, the simple model will be presented without reference to the work of Dalton, though he is mentioned in the independent practice, while a science story worksheet on the life and work of Dalton can be found in the additional materials on Boost.

Knowledge check

Display the *General science quiz*. There is no prior knowledge check in this part of the unit.

Guided explanation

What are atoms?

Introduce the topic by stating that everything is made of atoms – the furniture in the room, the water that comes from the taps, the air in the room and even us. Everything is made of atoms. Draw the following diagram to give students a sense of scale. Construct the diagram sequentially, with each step accompanied by an explanation, similar to the one that follows.

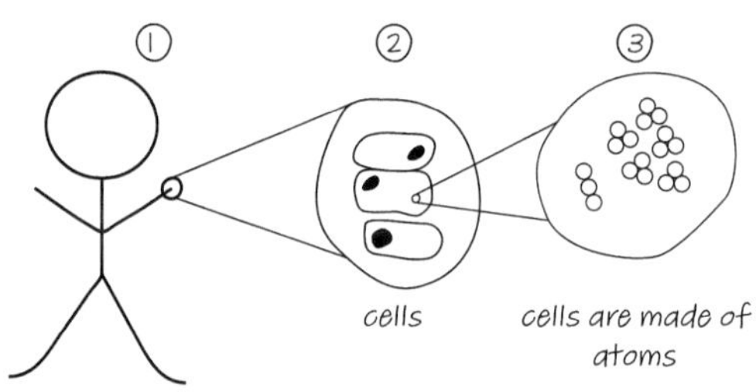

1 Say that we have already seen that we are made of cells.

2 Tell students that the human body contains billions of cells. These are too small to see by eye, but we can see them with a microscope.
3 Say that each cell is made of billions and billions of atoms. These are so small that we cannot even see them with the most powerful microscope.

Explain that, despite the incredible range of different materials like water, furniture, the air and living creatures, there are only 92 types of atom that exist naturally, and they make up everything. Atoms are the smallest unit of matter that can exist on its own. The names of each of the types of atom are listed on the Periodic Table of elements.

1 Display the Periodic Table on the second slide and give students time to look at it.
2 Call their attention back to you and use specific examples to explain that each box has the name of one type of atom in it. Each type of element also has some letters making up its symbol, but more on that later.
3 Tell students that if a substance is made up of all the same type of atom, then it is called an *element*.
4 Point out that the Periodic Table is not simply a list – it has a special layout. It is arranged into groups (vertical columns) and periods (horizontal rows), which have certain things in common.

Check for understanding

The Check for understanding (CFU) is best conducted using MWBs. Display the *CFU* slide and ask questions like the ones below. Tailor and adapt your questions to the class in front of you.

- What are all substances made of?
- What is the smallest unit of matter that can exist on its own? *If knowledge of protons, neutrons and electrons is unearthed or offered by students at this stage, say these are subatomic particles that you will discuss soon, but these particles cannot exist naturally on their own.*
- What is an element? *This is an important question and it is important that students know this answer almost verbatim. If many students get this wrong, ask it again at the end of the CFU.*
- How many different types of naturally occurring atom are there?
- Where are all the different types of element listed? *Shortening the answer to 'Periodic Table' is fine, but 'of elements' helps students remember what is listed on it.*
- What is a column on the Periodic Table called?
- What is a row on the Periodic Table called?

Independent practice review

Students practise independently by answering questions 1–8 from their Practice Books (page 45). Full answers can be found on the *IP answers* slides or in the answer documents on Boost. Notes are only included for questions, where useful.

Q	Notes
1, 7	Different ways of stating that everything is made of atoms, which are the smallest unit of matter.
3–5	Students should be able to use the correct language to describe the Periodic Table.
6	It is vital that elements are defined as made of only one type of atom, not by their presence on the Periodic Table, otherwise a misconception can arise that substances such as graphite or diamond are not elements.

Guided explanation

What is the structure of an atom?

Construct the diagram sequentially, with each step accompanied by an explanation, similar to the one that follows.

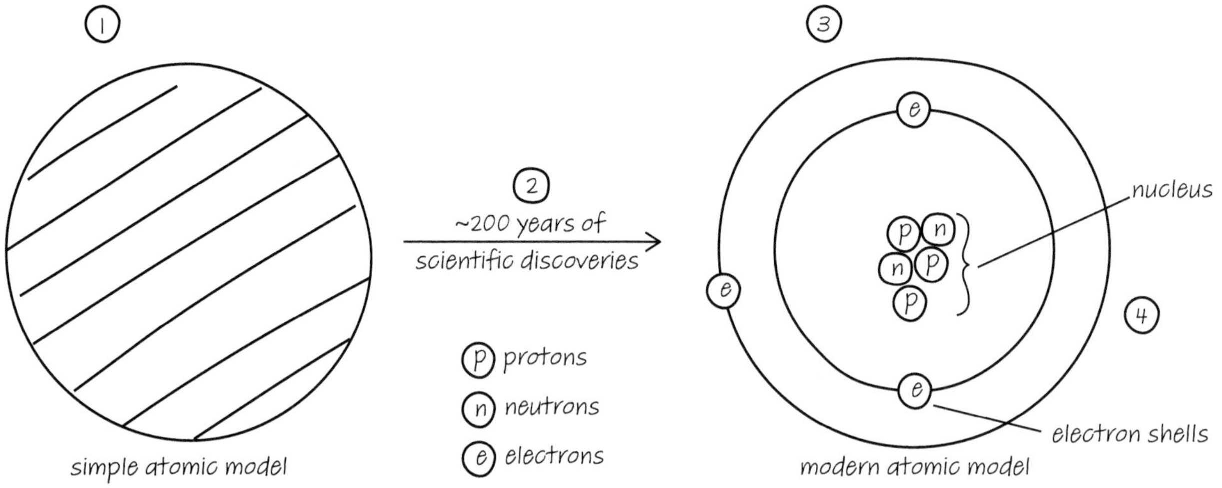

1 Originally scientists believed that, because atoms were the smallest thing, there was nothing smaller than atoms. These were tiny, indivisible particles.

2 Over the next 200 years, several scientific discoveries would show us that this simple model of an indivisible sphere was incorrect, since atoms are made up of three smaller particles. These smaller particles are called *protons, neutrons* and *electrons*.

3 (Draw the protons, neutrons and electrons in an atom, but do not label the nucleus or shells yet.) Because protons, neutrons and electrons are smaller than the level of the atom, we call them *subatomic particles*. However, since these subatomic particles cannot exist easily without each other, we call the *atom* the smallest unit of matter that can exist on its own. (If students are unsure of the term 'matter', tell them it just means 'things made of particles'.)

4 (Label the nucleus and electron shells.) The protons and neutrons are found at the centre of the atom, which we call the *nucleus*, whilst the electrons whizz around in the rest of the atom in electron shells. (Make it clear that these shells are not 'real' things like solid rings or tubes; they are just the routes the electrons take through the empty space.)

Check for understanding

Using MWBs, ask questions like the ones below.

- What are the names of the three subatomic particles?
- Where is the nucleus of an atom? *This may prompt questions from students about why this same word is used for atoms and cells. It is worth pointing out that the word means 'centre'.*
- Which subatomic particles are found in the nucleus?
- Which subatomic particles are found in shells around the nucleus?
- What is between the shells and the nucleus?
- What is between the electrons in the shells? *These two questions may prompt students to say 'air', but they should be corrected to understand there is nothing but empty space in most of the atom.*
- Display the *CFU* slide and draw an unlabelled atom. Ask students to name various parts.
- Draw the atom again and make some mistakes (e.g. putting protons in the shells). Ask students to identify the mistakes.

Independent practice review

Student Practice Book questions 9–18 (page 45).

Q	Notes
11	This *Because-But-So* writing frame question is designed to encourage students to think about how our model and understanding of the atom have changed with the discovery of subatomic particles.
18	This question encourages students to consider that atoms are not solid balls, and that much of the atom is empty space.

Guided explanation

How do we calculate the numbers of protons, neutrons and electrons in an atom?

Display the first slide and explain that the different types of atom listed on the Periodic Table differ by the number of protons, neutrons and electrons found inside each atom. The two numbers on the Periodic Table for each atom can be used to work out how many protons, neutrons and electrons each type of atom has. These two numbers are called the *atomic number* and the *atomic mass*.

On the second slide, illustrate this using lithium as an example, as shown below, accompanied by an explanation, similar to the one that follows.

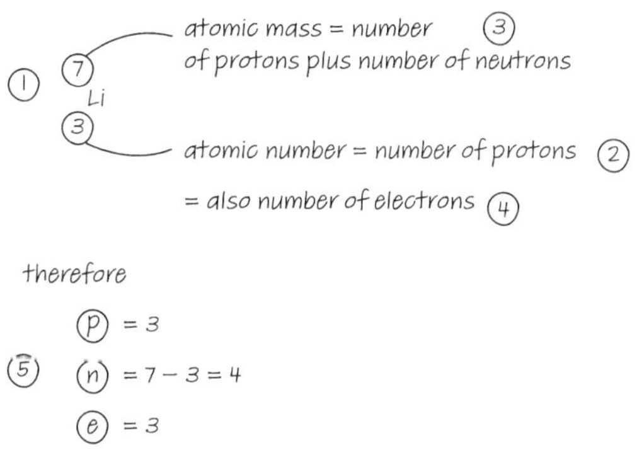

1 Using the first slide, show students that the numbers for atomic mass and atomic number have come directly from the Periodic Table.

2 On the second slide, circle and label the atomic number. Tell students that this is the number of protons in the atom.

3 Circle and label the atomic mass. Tell students that this is the number of protons and neutrons in the atom added together.

4 Tell students that the number of protons and electrons in atoms is the same, so the atomic number is also the number of electrons. (They do not need to know why this is the case as we are not introducing charge at this stage.)

5 Then go through how to use the atomic number and atomic mass to calculate the number of protons, neutrons and electrons in the lithium atom. Make sure you are explicit about the sum required to calculate the number of neutrons.

It is useful to begin with examples of elements that do not have the same number of protons and neutrons, so that there is no chance of a misconception taking hold that the numbers of protons, neutrons and electrons are all the same.

Note: It is unhelpful to use terms such 'top number' and 'bottom number' when explaining nuclide notation, since different versions of the Periodic Table change the position of these numbers. Highlight explicitly that all versions of the Periodic Table contain a key that shows which number is the atomic number and which is the atomic mass.

Work through another three examples, such as aluminium, magnesium and phosphorus, as shown below. As you progress through the examples, increasingly involve the students via questioning.

aluminium

27 Al 13

atomic mass = (P) + (n)

atomic number = (P) = (e)

(P) = 13

(n) = 27 − 13 = 14

(e) = 13

magnesium

24 Mg 12

atomic mass = (P) + (n)

atomic number = (P) = (e)

(P) = 12

(n) = 24 − 12 = 12

(e) = 12

phosphorus

31 P 15

atomic mass = (P) + (n)

atomic number = (P) = (e)

(P) = 15

(n) = 31 − 15 = 16

(e) = 15

Be sure to explicitly show students that the atomic mass and atomic number have been copied directly from the Periodic Table on each occasion. Repeat the same steps each time.

1. Circle and label the atomic number. Show that this is the number of protons.
2. Circle and label the atomic mass. Show that this is the number of protons plus neutrons.
3. Add that the atomic number is also the number of electrons.
4. List the number of protons, neutrons and electrons in each atom, with the calculation for neutrons shown explicitly.

Check for understanding

Display the first *CFU* slide (for beryllium) and, using MWBs, ask questions like the ones below.

- What is the atomic number?
- What is the atomic mass?
- How many protons does beryllium have?
- How many neutrons does beryllium have?
- How many electrons does beryllium have?

Begin by asking students about each type of subatomic particle separately. If they show good understanding, for further examples you can ask them about all the subatomic particles simultaneously. But do not do this until students are demonstrating good understanding of each separately.

Display the second and third slides with different atoms and ask similar questions.

Display the fourth slide, for hydrogen, and ask similar questions.

Hydrogen is unusual in that it has 0 neutrons. It is a good choice for your final CFU example because students will be very confident in their understanding of the application of the rules if they write 0 neutrons. Students may ask how an atom can have 0 neutrons, in which case explain that hydrogen is a special case because it is so small.

Independent practice review

Student Practice Book questions 19–25 (pages 46–47).

Q	Notes
20	Students will need a Periodic Table to find the atomic number and atomic mass of some elements.
21–25	These questions encourage students to work backwards and identify atoms, having been given their subatomic particles and going back to the Periodic Table.

How many electrons can go in each shell?

Recap that we have seen that electrons are found in shells orbiting the nucleus. Explain that there is a maximum number of electrons that each shell can hold – e.g. the first shell can hold a maximum of two electrons, the second shell a maximum of eight electrons and the third shell can also hold eight electrons.

Draw the diagram below sequentially, with each step accompanied by an explanation, similar to the one that follows.

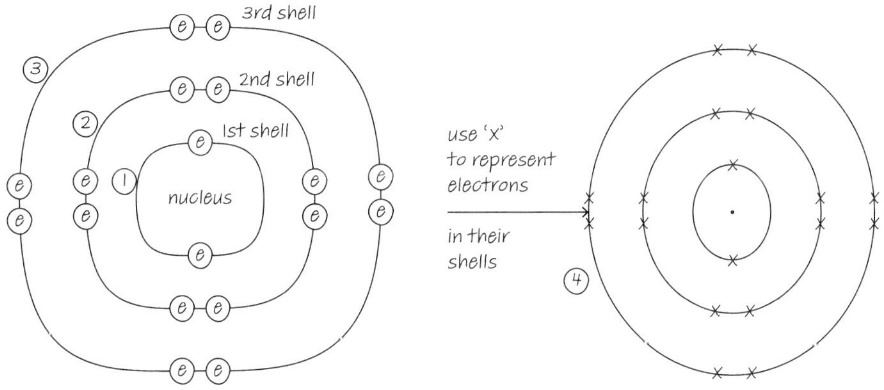

1 The nucleus is very small compared with the size of the whole atom – remember that most of the atom is just empty space – so we draw it as a dot because now we are talking about electron shells. The first shell of electrons can hold up to two electrons. Remember that this is not a ring or tube; it is just the space that the electrons travel through. (Label the '1st shell'.)

2 The second shell of electrons can hold up to eight electrons. This shell is further away from the nucleus, and we count the shells from the inside to the outside. (Label the '2nd shell'.)

3 The third shell can also hold up to eight electrons. This shell is further away again from the nucleus. (Label the '3rd shell'.)

4 Drawing a circle with an 'e' in it takes some time, especially when we have a lot of electrons in an atom. We can use any symbol to represent electrons, but scientists often use 'X', as these are quick and easy to draw. Therefore, the same atom with 18 electrons in three shells can be redrawn more quickly like this.

Explain that the shells that are closest to the nucleus are filled with electrons first, so there need to be two electrons in the first shell before any electrons are found in the second shell.

Similarly, the second shell must be full before any electrons are put in the third shell.

Draw the diagram below to explain how electrons are arranged in atoms. Construct the diagram sequentially, with each step accompanied by an explanation, similar to the one that follows.

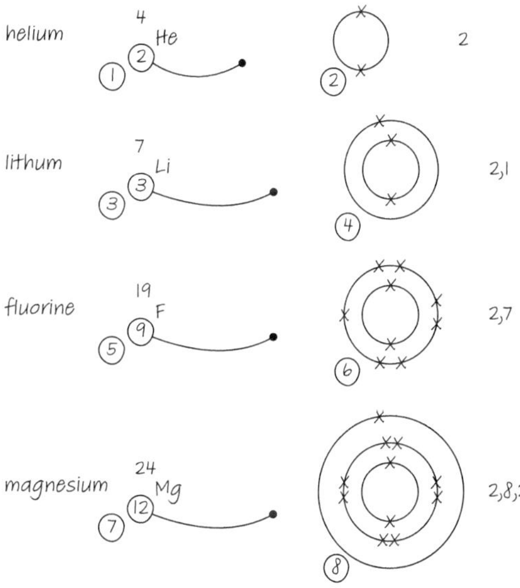

1 Take helium as our first example. It has an atomic mass of 4 and an atomic number of 2. This means it has two electrons.

2 Those electrons can both fit in the first shell and no electrons are left over for the next shell. Its electron configuration is 2.

3 Lithium is in Group 1 on the second period of the Periodic Table. It has an atomic mass of 7 and an atomic number of 3. This means it has three electrons.

4 Those electrons cannot all fit in the first shell, which can only hold two electrons. The third, remaining electron goes into the second shell. This means the electron configuration is 2,1. Lithium has two shells of electrons with one electron in the outer shell.

As you progress through the next examples, increasingly involve students via questioning.

5 Fluorine is in Group 7 on the second period of the Periodic Table. It has an atomic mass of 19 and an atomic number of 9. This means it has nine electrons.

6 Two of those nine electrons fit in the first shell, and the remaining seven electrons go in the second shell. Its electron configuration is 2,7. Fluorine has two shells of electrons with seven electrons in the outer shell.

7 Magnesium is in Group 2 on the third period of the Periodic Table. It has an atomic mass of 24 and an atomic number of 12. This means it has 12 electrons.

8 Two of those 12 electrons fit in the first shell and eight can fit in the second shell. This leaves two more electrons, which must go in the third shell. Its electron configuration is 2,8,2. Magnesium has three shells of electrons with two electrons in the outer shell.

Check for understanding

Using MWBs, ask questions like the ones below.

- Which subatomic particles are found in the shells?
- Up to how many electrons can the first shell hold?
- Up to how many electrons can the second shell hold?
- Up to how many electrons can the third shell hold?

Any questions relating to the fourth and later shells are best left for the moment, although if a question is asked, 'more than 8' is sufficient for the time being.

Ask some questions that require students to apply the rules to generic atoms, like the ones below.

- An atom has four electrons. Draw how they would be arranged in an atom. Use 'X' to represent each electron. *Check that students are drawing two electrons in the first (inside) shell and two electrons in the outer shell.*
- An atom has seven electrons. Draw how they would be arranged in an atom. Use 'X' to represent each electron. *Whilst the five electrons in the outer shell do not need to be spread out at top, bottom, left and right, it is useful to encourage students to do this.*
- An atom has 14 electrons. Draw how they would be arranged in an atom. Use 'X' to represent each electron.

Ask some questions about specific atoms, like the ones below.

- Display the first *CFU* slide, which shows the nuclide notation for carbon (C). Ask students to draw the electron arrangement without saying which number they should use.
- Display the second slide, which shows the nuclide notation for aluminium (Al). Ask students to draw the electron arrangement without saying which number they should use.

The independent practice for this section is included in the questions that follow the guided explanation below.

Guided explanation

How is electron arrangement linked to the Periodic Table?

Start this explanation by looking at four elements from Period 2. Display the first slide and click through each element sequentially, accompanying each one with an explanation, similar to the one that follows.

1 First check that students can find and name each element on the slide.
2 Then ask them to check that the electron diagram is correct.
3 Ask them to suggest how the group number is linked to the electron configurations.
4 When the correct answer is identified, confirm to the whole class that the group number is the same as the number of electrons in the outer shell of electrons. This is true for the whole Periodic Table.
5 Repeat this process for the second slide, which shows elements in the same group but different periods. Ask students to suggest how the period number is linked to the electron configurations.
6 When the correct answer is identified, confirm to the whole class that the period number is the same as the number of electron shells. Display the third slide and say this is true for the whole Periodic Table.

Check for understanding

Display the Periodic Table on the *CFU* slide and, using MWBs, ask questions like the ones below.

- Bromine is in Group 7. How many electrons are in the outer shell of a bromine atom?
- Caesium is in Group 1. How many electrons are in the outer shell of a caesium atom?
- Barium is in the sixth period of the Periodic Table. How many shells of electrons does a barium atom have?
- Gallium is in the fourth period of the Periodic Table. How many shells of electrons does a gallium atom have?
- Find iodine on the Periodic Table. How many shells of electrons does it have? How many electrons does it have in its outer shell?
- Find barium on the Periodic Table. How many shells of electrons does it have? How many electrons does it have in its outer shell?
- Nitrogen has seven electrons. Draw how they will be arranged in a nitrogen atom. Use Xs for electrons.
- Magnesium has 12 electrons. Draw how they will be arranged in a magnesium atom. Use Xs for electrons.

Independent practice review

Student Practice Book questions 26–43 (pages 47–48).

Q	Notes
39	It is important that students are including Period 1 (H and He) when counting the periods on the Periodic Table.
40, 41	These questions pull together students' knowledge of the modern atomic structure. They should be able to deduce the numbers of all the subatomic particles and the electron configuration from the Periodic Table.
43	This open writing question introduces John Dalton as an important scientist, and allows students to compare their knowledge of the modern atomic model with the simple atomic model theorised by John Dalton.

C2.2 Symbols and formulae

Students' learning objective: *I am learning about chemical symbols and chemical formulae so that I can understand the way chemists talk about atoms and chemicals.*

The aim of this topic is to introduce students to the symbols and formulae used in chemistry and what they represent.

Knowledge check

After working through the *General science quiz*, display the *Knowledge check* slide. The questions are best delivered one by one on mini-whiteboards (MWBs). If students do not know the answer to any of them, you should reteach the relevant concept. Answers are not on the slides because it is best to go over them together as a class.

Questions 4 and 5 are to gauge understanding of knowledge from KS2 and outside school.

	Question	Answer	Notes
1	What is an atom?	The smallest unit of matter that can exist on its own	
2	Where are all the different types of atom listed?	The Periodic Table of elements	
3	How many different types of atom are there?	92 (naturally occurring)	Around 100 or 118 are also acceptable answers.
4	If a fruit bowl has six apples and three oranges in it, what is the ratio of apples to oranges?	2:1 (two apples for every one orange)	This is a KS2 maths topic, so most students should be familiar with the concept but may need reminding or checking.
5	What is H_2O?	Water	Many students will know the chemical formula of water. This question is a good one for them to refer to later when they forget what a *chemical formula* is.

Guided explanation

What are chemical symbols?

This explanation will teach students what chemical symbols are – a system of one or two letters, used everywhere in the world, so that scientists have a quick way to explain which atoms they are talking about.

Display each of the slides in turn, with each one accompanied by an explanation, similar to the one that follows.

1 (Display the Periodic Table on the first slide.) There are 92 naturally occurring types of atom and 26 more have been made in the laboratory (so far). The current Periodic Table therefore lists 118 different types of atom. Note that some Periodic Tables, such as the one in your Knowledge Book, do not have the lanthanide and actinide elements, which are quite rare, to save space.

These 92 types of atom make up everything in the world around us, from the Sun to the air that we breathe, the furniture in this room and us. Everything is made from these 92 types of atom.

2 (Display the pictures of gold and tungsten on the second slide.) Often when these different types of atom have been known about for a long time, they have different names in different languages all around the world. For example, gold is called *oro* in Spanish and Italian, while tungsten is called *Wolfram* in German. In order for scientists all over the world to be able to communicate clearly, it was decided that each type of atom should have a universal chemical symbol that everyone would use, even if the names were different.

3 (Display the Periodic Table again.) It was decided that one or two letters should be used to represent each type of atom – this is called its *chemical symbol*. As you can see from the Periodic Table the chemical symbols take the form of either one capital letter or one capital letter and one lower-case letter.

It would be useful to highlight some of the chemical symbols and write them out for the class, explicitly highlighting the capital and lower-case letters. For example:

- Hydrogen has the symbol 'capital H'.
- Helium has the symbol 'capital H, lower-case e'.
- Lithium has the symbol 'capital L, lower-case i'.
- Beryllium has the symbol 'capital B, lower-case e'.
- Boron has the symbol 'capital B'.

Point out that many of the symbols are related to the English names of the types of atom, but as we shall see that is not always the case. For example:

- Gold has the symbol 'capital A, lower-case u' (which is from its Latin name *aurum*, which is related to *oro* in Italian and Spanish).
- Tungsten has the symbol 'capital W' (which is from its German name, *Wolfram*).

Check for understanding

Display the Periodic Table on the *CFU* slide and ensure that students also have access to a copy in their exercise books or in the Knowledge Book. Using MWBs, ask questions like the ones below.

- Looking at the second period of the Periodic Table, what is the chemical symbol for carbon?
- Looking at the third period of the Periodic Table, what is the chemical symbol for sodium? *You can tell students that this symbol comes from the Latin name for sodium, natrium, because it is from the natural (sea) salt.*
- Looking at the third period of the Periodic Table, what is the chemical symbol for chlorine? *Explicitly highlight to students that this symbol has a lower-case 'l' and not a capital 'I', which appear similar in many fonts.*
- Looking at the fourth period of the Periodic Table, what is the chemical symbol for potassium? *You can tell students that this symbol comes from the Latin name for potassium, kalium.*
- Looking at the fourth period of the Periodic Table, what is the chemical symbol for copper? *This is an important one to highlight because the name comes from the Latin cupric rather than the English name, so the symbol is not Co, which is cobalt.*
- A student has written the chemical symbol for beryllium as 'BE' (*write this on the board for students to see*). Why is this incorrect?
- Looking at your Periodic Table, name the element with the symbol V.
- Looking at your Periodic Table, name the element with the symbol P.
- Find the element in Period 4, Group 5. What is its name? What is its symbol?
- Find the element in Group 1, Period 5. What is its name? What is its symbol?

Independent practice review

Student Practice Book questions 44–48 (pages 48–49).

Q	Notes
45	This question helps students become familiar with the Periodic Table, especially the first four periods, and with getting to know the names and chemical symbols of different types of common atom.
46	This question uses non-examples to highlight common errors that students make in writing chemical symbols.

Guided explanation

What are chemical formulae?

Begin by saying that very few substances are made of just one atom, and that most substances are made of lots of atoms joined together. The chemical formula tells us what ratios of atom types are within a substance. We use a combination of chemical symbols and numbers in the chemical formula.

Draw the diagram below, with each step accompanied by an explanation, similar to the one that follows.

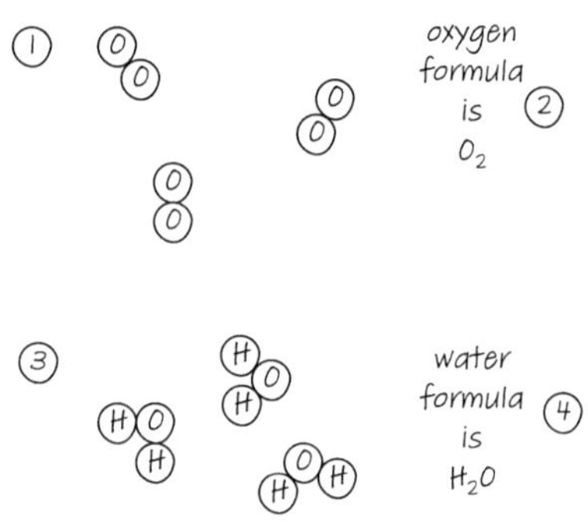

1 When a substance is made of particles that are a few atoms joined together, we call it a *molecule*. We count how many of each type of atom are bonded together in the molecule. For example, oxygen gas is made from molecules formed of two oxygen atoms chemically joined together.

2 We show that each molecule is made from two oxygen atoms in the formula by writing 'O_2'. It is important that the '2' is subscript (smaller and lower down than the O) and to the right of the O.

3 Similarly, you may already know that the formula of water is H_2O. This is because water particles are molecules made from two hydrogen atoms and a single oxygen atom joined together.

4 This time the '2' is subscript again and to the right of the 'H', because there are two hydrogen atoms. The ratio of hydrogen to oxygen is 2:1.

Explain that not all substances are made of molecules. Some substances are made of billions of atoms all joined together.

Draw the diagram below to show how we represent the formulae of these substances slightly differently. Accompany each step with an explanation, similar to the one that follows.

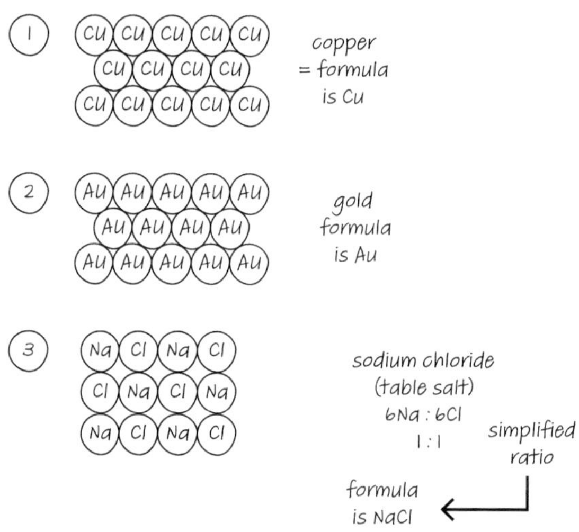

1 When all of the atoms are of the same type, we do not count how many atoms there are; we just say that they are all the same. For example, a piece of copper is made from billions of copper atoms, and we say that its chemical formula is Cu. (Point out that this type of diagram, where we draw all the atoms as circles, is called a *bubble diagram*.)

2 Similarly, a piece of gold is made from billions of gold atoms, and we say that its formula is Au, since this is the chemical symbol for gold atoms.
3 When different types of atom combine to form substances, they always do so in fixed whole-number ratios. For example, table salt is made from billions of sodium atoms and billions of chlorine atoms in a structure like this. If we count how many chlorine and sodium atoms are in this substance, we can see that there are six of each. This ratio can be simplified to 1:1, and thus the chemical formula of table salt is NaCl.

Check for understanding

Display the first *CFU* slide and click to reveal the name of each substance in turn. Using MWBs, ask questions like the ones below.

Ensure you are clear with students when you are asking for chemical symbols, formulae or diagrams. You may want to write down exactly what you are looking for too.

- What is the chemical formula of water? *Check that the letters are capitals and that the number is subscript (below the level of the cross bar of the 'H').*
- How many H atoms are there in a water particle?
- How many O atoms are there in a water particle?
- Draw a bubble diagram for a water particle. *The order of the bonding does not matter at this stage; it is only important that there is one O atom and two H atoms.*
- What is the chemical formula of oxygen gas? *Check that students are not confusing chemical symbol and chemical formula.*
- Draw a bubble diagram for oxygen gas.
- What is the chemical symbol for hydrogen atoms?
- Hydrogen gas has particles that are made from two hydrogen atoms joined together. What will its chemical formula be?
- Hydrogen peroxide is a chemical with the formula H_2O_2. Write this formula on the board. How many different types of atom are there in hydrogen peroxide? *Check that students understand that the symbols show the different types of atom.*
- Hydrogen peroxide is a chemical with the formula H_2O_2. How many atoms are there in total in hydrogen peroxide? *Check that students understand how to calculate the total number.*
- Draw a bubble diagram for hydrogen peroxide (H_2O_2). *The order of the bonding does not matter at this stage; it is only important that there are two O atoms and two H atoms.*

Display the second slide, which has a bubble diagram of MgO, and ask students to write the formula of the substance shown.

Check that students understand that if a solid is made from lots of particles, then the formula is the simplest ratio, so in this case, with xMg and xO, it is MgO as the simplest ratio, which is 1:1. The chemical formula is not Mg_xO_x, where x is the number of particles of Mg and O drawn.

Display the third slide, which has bubble diagram of Mg, and ask students to write the formula of the substance shown.

Check that students understand that if a solid is made from lots of particles, then the formula is the simplest ratio, so in this case it is Mg, not Mg_x (where x is the number of particles drawn).

Display the fourth slide, which has bubble diagrams of CO_2, and ask questions like the ones below.

- Write the formula of the substance shown. *Correct any students who give an answer with a multiple of CO_2, e.g. C_2O_4.*

- What is a molecule?
- How can you tell that carbon dioxide is a molecule?

If students have struggled with these examples, draw more similar ones and do further CFU.

Independent practice review

Student Practice Book questions 49–63 (pages 49–51).

Q	Notes
56–60	Students work with unfamiliar substances and formulae. For the bubble diagrams the order of the bonding is not important – it is just important that the correct numbers and types of atom are in the substances. Some students may find it easier to start with part d each time and draw the bubble diagram first to help visualise the molecule.
61, 63	Students work with unfamiliar substances and formulae, drawing bubble diagrams from formulae and formulae from bubble diagrams.

C2.3 Elements and compounds

Students' learning objective: *I am learning about elements and compounds so I can explain the difference between them.*

The aim of this topic is to introduce students to compounds and the idea that they have properties that are completely different from the individual elements that form the compound.

Knowledge check

After working through the *General science quiz*, display the *Knowledge check* slides and, using MWBs, deliver the questions one by one, reteaching concepts as necessary.

	Question	Answer	Notes
1	What is an element?	A substance made of only one type of atom	'Found on the Periodic Table' is insufficient.
2	How many types of atom are there?	92 (naturally occurring)	
3	Where are the different types of elements listed?	The Periodic Table of elements	'Periodic Table' is sufficient.
4	What is the name for the one or two letters that are used to represent a type of atom?	Chemical symbol	
5	What form do chemical symbols take?	One capital letter or one capital letter and one lower-case letter	
6	What do we call the chemical symbols and numbers that tell us how many of each type of atom are in a substance?	Chemical formula	

	Question	Answer	Notes
In the following questions the formulae should be displayed.			
7	How many different elements are there in are H_2?	One	These are two atoms of the same element joined in a hydrogen molecule.
8	How many different elements are there in NaF?	Two	Check that students recall that the small 'a' is part of the symbol for sodium.
9	How many different atoms are there in CO_2?	Two	Not three – the question asks how many *different* atoms.
10	How many different atoms are there in H_2SO_4?	Three	S and O are both capital letters, representing two different elements.
11	What are the three states of matter?	Solid, liquid and gas	

Guided explanation

What are elements and compounds?

Recap that we have seen that there are 92 types of atom listed on the Periodic Table. If a substance is made of only one type of atom, then it is an element. When we looked at the formulae of substances, we saw that many substances, such as water, were made from more than one type of atom chemically joined together. We call these substances *compounds*. Whereas there are only around 100 elements, there are hundreds of thousands of combinations, which can form different compounds.

Draw the diagram below to illustrate that all chemicals are either elements or compounds, and there is no overlap.

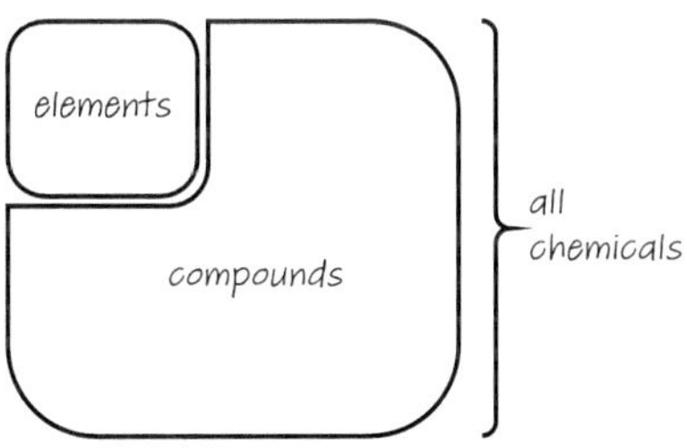

Explain that since a compound is made of more than one type of atom chemically joined, and elements are made entirely of only one type of element, it means that all chemicals fit into one of these groups and there is no overlap at all. Say that it is usually easy to tell if a substance is an element or a compound, especially if you have its chemical formula, because compounds will have more than one chemical symbol in them.

Draw the diagram below to exemplify elements and compounds, constructing it sequentially, with each step accompanied by an explanation, similar to the one that follows.

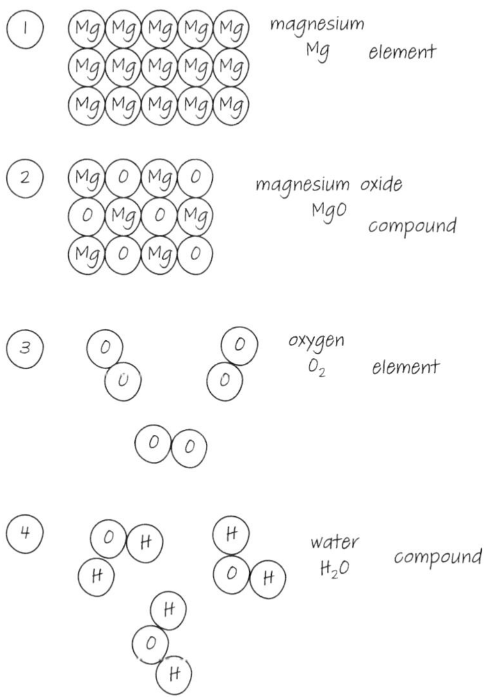

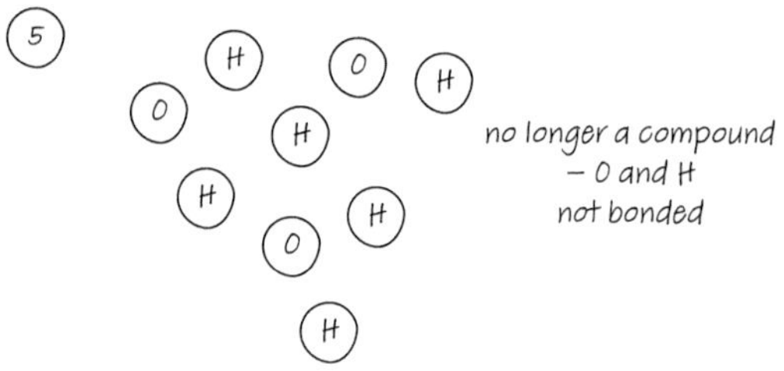

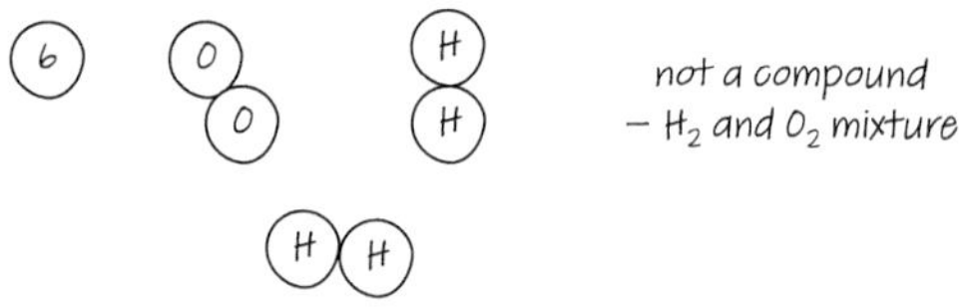

1 Taking magnesium as an example, it has the formula Mg, which only has one symbol and is therefore an element. As we can see from its bubble diagram, it is only made of Mg atoms, which again shows us that it is an element.

2 Magnesium oxide has the formula MgO, which has two symbols and is therefore a compound made from magnesium and oxygen atoms. We can also see from the bubble diagram that this substance is made from both magnesium and oxygen atoms joined together, and is therefore a compound.

3 Oxygen gas has the formula O_2, which shows us that oxygen particles are molecules made of two oxygen atoms joined together. Even when there are lots of oxygen molecules, they are all oxygen atoms and thus oxygen gas is an element made of only one type of particle.

4 Water has the formula H_2O, which shows us that water particles are molecules made of two hydrogen atoms and one oxygen atom. Even when there are lots of water molecules, this ratio is constant. Because water is made of oxygen and hydrogen atoms bonded together, it is a compound.

5 If we were to split all the H and O atoms apart from each other, is this a compound? No. This would no longer be a compound because the H and O atoms are not bonded to each other.

6 If they were to bond only to atoms the same as them, is this a compound? No. This is a mixture of H_2 and O_2 – two elements.

Check for understanding

Display the first *CFU* slide and, using MWBs, ask questions like the ones below.

- What is an element?
- What is a compound? *These two questions are vital and should be asked again in the CFU sequence if some students did not understand at the start.*

Formulae for the following questions should be written out on the board so their symbols are clear.

- Carbon dioxide has the formula CO_2; is it an element or a compound?
- Methane has the formula CH_4; is it an element or a compound? Explain why. *Since these 'element or compound?' questions could be a 50:50 guess, students should write brief explanations of their answers. Students not writing explanations should be followed up with a Cold Call to explain their answer.*
- Potassium has the formula K; is it an element or a compound? Explain why.
- Fluorine has the formula F_2; is it an element or a compound? Explain why.
- Hydrogen fluoride has the formula HF; is it an element or a compound? Explain why.

Display the second slide with the four bubble diagrams, clicking to reveal each one in turn. Ask students whether the diagrams represent elements or compounds. (You could also ask students to identify the formulae of these substances to check their skill of identifying formulae from the last topic.)

Independent practice review

Student Practice Book questions 64–75 (pages 52–53).

Q	Notes
66, 67	These questions confirm students' understanding of the definitions of elements and compounds.
71, 75	The questions allow students to practise classifying substances as elements or compounds, based on bubble diagrams or formulae.

Guided explanation

How do we name compounds?

Begin by explaining that a few compounds have special names, such as water (H_2O) and methane (CH_4); however, these are the only two that we need to learn at this stage. Most compounds are simply named using the elements that are combined within the compound.

Tell students that there are four simple rules that we use to name compounds systematically. Illustrate this by writing some examples on the board, step by step, as shown below.

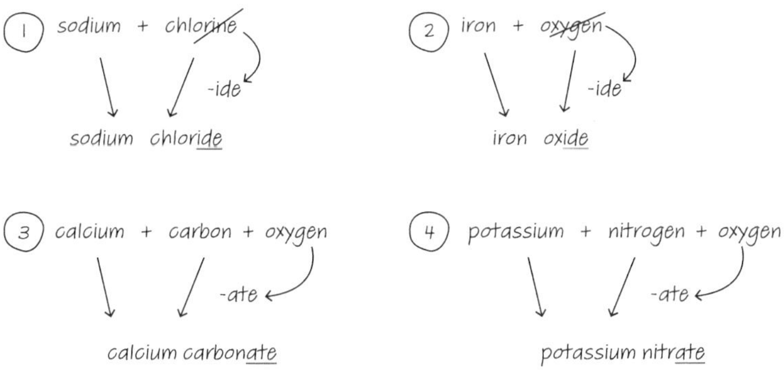

1 **Rule 1:** The element further to the left in the Periodic Table (usually a metal) comes first in the name of the compound. For example, when sodium and chlorine react together, sodium is further to the left on the Periodic Table so comes first.
   Rule 2: The second element in the name changes the end of its name to '-ide' to show that it is bound to the first element in a compound. When sodium and chlorine react together, chlorine changes to chloride. The compound is named sodium chloride.
2 Looking at the reaction of iron and oxygen, iron is further to the left on the Periodic Table so comes first, and oxygen changes to oxide. Therefore, the compound is iron oxide.
3 **Rule 3:** If there are three elements in the compound and one of them is oxygen, the second element in the name changes its ending to '-ate' instead of '-ide' to show that oxygen is the third element. For example, when calcium, carbon and oxygen atoms combine in a compound, calcium is the first name in the compound since it is furthest to the left, and carbon's ending is changed to '-ate' to show that oxygen is the third element within the compound. The compound's name is therefore calcium carbonate.
4 Looking at the reaction of oxygen, potassium and nitrogen combining to form a compound, potassium is the first element listed in the name because it is furthest to the left and the ending of nitrogen is changed to '-ate' because oxygen is also present. The compound's name is therefore potassium nitrate.

Check for understanding

Display each of the *CFU* slides in turn, clicking to reveal the pairs of elements or compound names. Using MWBs, ask questions like the ones below.

- What is the name of the compound formed when magnesium reacts with oxygen?
- What is the name of the compound formed when copper reacts with chlorine?
- What is the name of the compound formed when fluorine reacts with zinc? *Ask at least one question where the non-metal is listed first in the question.*
- (Write down 'calcium bromide'.) What types of atom are found in calcium bromide? *Students may need to use their Periodic Tables to look up the name 'bromine', but they should be encouraged to do so.*
- What is the name of the compound formed when magnesium, carbon and oxygen bond together?
- What types of atom are found within the compound copper sulfate?
- What is the name of the compound formed by calcium, sulfur and oxygen?

The independent practice for this section is included in the questions that follow the guided explanation below.

Guided explanation

How do we name compounds? *(continued)*

Illustrate the fourth rule by drawing the diagram below sequentially, with each step accompanied by an explanation, similar to the one that follows.

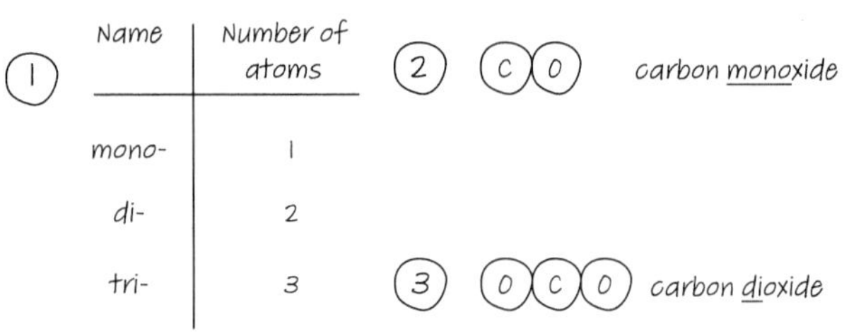

1 **Rule 4:** This rule is used when the same types of atom can form different compounds when bonded together in different ratios. In this case, the words *mono*, *di* and *tri* are used to denote 1, 2 and 3 atoms, respectively.

2 (Draw examples of compounds with multiple oxides.) For example, carbon can bond with oxygen in the ratio 1:1. This compound is called carbon monoxide (CO).

3 Carbon can also bond with oxygen in the ratio 1:2. This compound is called carbon dioxide (CO_2).

Check for understanding

Display the *CFU* slide and click to reveal each compound in turn. Say that nitrogen can form compounds with oxygen in different ratios. Then, using MWBs, ask questions like the ones below.

- What is the compound with the formula NO? *'Mononitrogen monoxide', whilst incorrect, should be noted as correct and logical use of the rules. 'Oxygen mononitride' has used rule 1 incorrectly.*
- What is the compound with the formula NO_2? *'Mononitrogen dioxide', whilst incorrect, should be praised as correct and logical use of the rules. 'Dioxygen nitride' has used rule 1 incorrectly.*

Independent practice review

Student Practice Book questions 76–83 (pages 53–54).

Q	Notes
76, 77	These two questions check that students have remembered the names of the two key compounds that do not follow the compound naming rules.
78, 79, 82	These questions check that students can use the naming rules correctly to identify the atoms in compounds, and the ratios of different atoms within them.
83	Hydroxide compounds do not follow the naming rules and students are introduced to them in this question, but they are not yet expected to learn the formula or name of hydroxides.

Guided explanation

What are chemical and physical properties?

Tell students that we are now going to compare compounds and the elements that form them. Draw the diagram below sequentially, with each step accompanied by an explanation, similar to the one that follows.

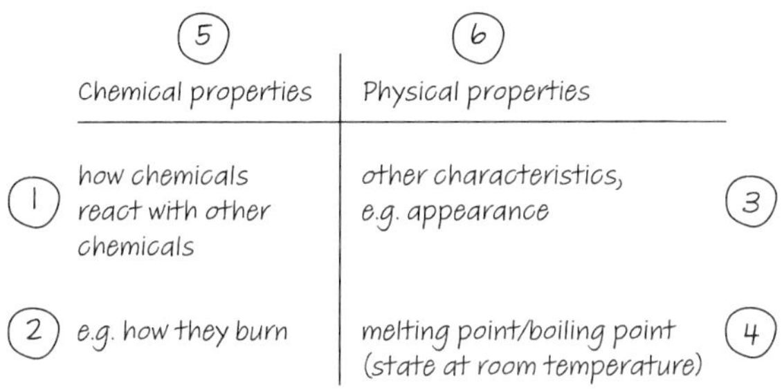

1 We can compare how substances react with other chemicals.

2 There are many chemical reactions we could compare, e.g. how easily a substance burns in oxygen. We are going to learn a lot more about chemical reactions in Year 8, so we will stop here for now.

3 We can also compare other characteristics of chemicals, e.g. their appearance (what they look like).

4 Other characteristics might include the state of the chemical at room temperature, which is set by its melting point and boiling point.

5 Collectively we call all the ways in which a chemical reacts with other chemicals its *chemical properties*.

6 We call all the other properties of a substance its *physical properties*.

Check for understanding

Display the *CFU* slide and, using MWBs, ask questions like the ones below.

- Gold is a soft metal that is easy to bend. Is this a chemical property or a physical property? Explain how you know.
- Give another physical property of gold. *Students may know its colour, that it is a solid or that it conducts electricity because it is metallic. Share good answers with the rest of the class.*
- Methane (from the gas taps) is very easy to burn in oxygen. Is this a chemical property or a physical property? Explain how you know.
- Give a physical property of methane. *Students may state that it is a colourless gas, with a smell (not worth correcting at this stage), that it can be compressed and that it can flow. Share good answers with the rest of the class.*
- Give a physical property of water. *Students may state that it is a colourless and transparent liquid, that its melting point is 0°C and its boiling point is 100°C, that it can flow, but it cannot be compressed.*

The independent practice for this section is included in the questions that follow the guided explanation below.

How are compounds related to the elements that form them?

Recap that compounds are formed when two or more different types of atom join together to form a new substance. However, the properties of the compounds that are formed are different from and unrelated to the elements that formed them.

Draw the diagram below, building up slowly and asking lots of questions to check understanding along the way. Accompany each step with an explanation, similar to the one that follows.

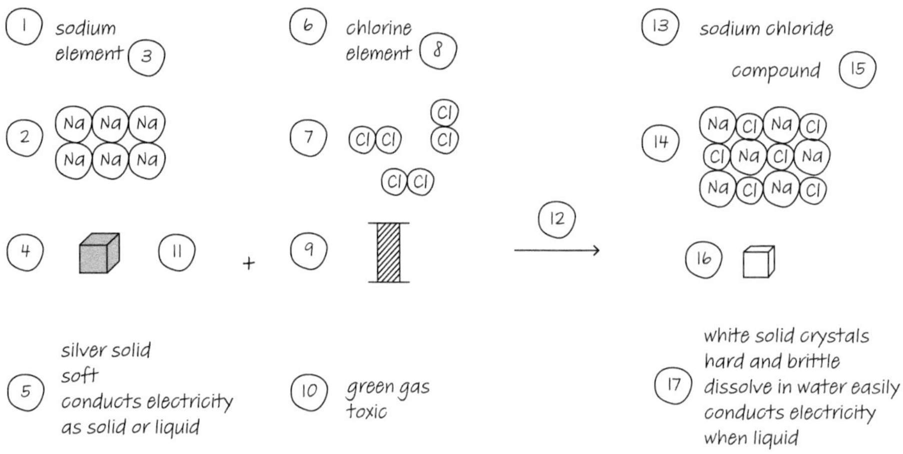

1 First let us look at sodium.

2 Sodium is made of lots of sodium atoms bonded together like this.

3 If all these atoms are the same type, then sodium is an element.

4 Sodium is often in cubes like this.

5 Sodium is a silver colour. It is soft and easy to bend and cut, and it conducts electricity as a solid or as a liquid.

6 Now let us look at a different chemical called chlorine.

7 Chlorine is made of chlorine molecules. Each molecule is two chlorine atoms chemically joined together.

8 Because chlorine is entirely made of chlorine atoms, it is also an element.

9 Chlorine is stored in gas jars or tanks because it is a gas.

10 It is a green gas, which is very toxic and poisonous.

11 We can react these two elements together.

12 When we do they produce a different substance.

13 This substance is called sodium chloride, but you know it as the table salt that we put on our food.

14 The atoms in sodium chloride are arranged alternately like this.

15 Because sodium chloride is made of sodium and chlorine atoms it is a compound, not an element.

16 You know that table salt is made of little grains, which are little cube-shaped crystals.

17 These crystals are white solids, which are hard and brittle, but they dissolve easily in water. Sodium chloride can conduct electricity, but not as a solid, only when it is a liquid.

Tell students that we can summarise these properties in a table, so we can easily compare them. Display the second slide. Click to reveal each cell sequentially, talking students slowly through the properties of sodium, chlorine and sodium chloride.

This table can also be found in Topic C2.3 of the Knowledge Book.

Make sure students understand that the physical and chemical properties of sodium chloride are completely different from the chemical and physical properties of the elements sodium and chlorine.

Check for understanding

Display the first *CFU* slide and, using MWBs, ask questions like the ones below.

- What are chemical properties?
- Give three examples of physical properties. *There are three listed in the Knowledge Book, but there are many acceptable correct answers.*
- What colour is sodium metal?
- What colour is chlorine gas?
- What colour is sodium chloride (table salt)?
- Is sodium an element or a compound? Explain how you know.
- Is sodium chloride an element or a compound? Explain how you know.
- Is chlorine an element or a compound? Explain how you know.
- What state is sodium at room temperature?
- What state is chlorine at room temperature?
- What state is sodium chloride (table salt)?
- Sodium reacts violently with water. Is this a chemical or a physical property? Explain how you know.
- How are the physical properties of compounds related to the physical properties of the elements that formed them?
- How are the chemical properties of compounds related to the chemical properties of the elements that formed them?

Display the second slide. Click to reveal each cell of the table sequentially, slowly talking students through the chemical and physical properties of hydrogen gas (H_2), oxygen gas (O_2) and water (H_2O). Explicitly highlight the differences between the compound water and the elements that form it. Then ask questions like the ones below.

- Give an example of a chemical property.
- Give an example of a physical property.
- Give an example of how water has different physical properties from oxygen and hydrogen.
- Give an example of how water has different chemical properties from oxygen and hydrogen.

Independent practice review

Student Practice Book questions 84–93 (pages 54–55).

Q	Notes
84–91	These questions check that students understand the basics of chemical and physical properties, and that compounds are different from the elements that form them, without students requiring in-depth knowledge of unusual chemicals.

Guided explanation

What are hazards and risks?

Tell students that before we do an investigation into the properties of elements and compounds, it is important to talk about how to stay safe when using dangerous chemicals, because this is the first time we have used some hazardous chemicals.

Draw the following diagram sequentially, with each step accompanied by an explanation, similar to the one that follows.

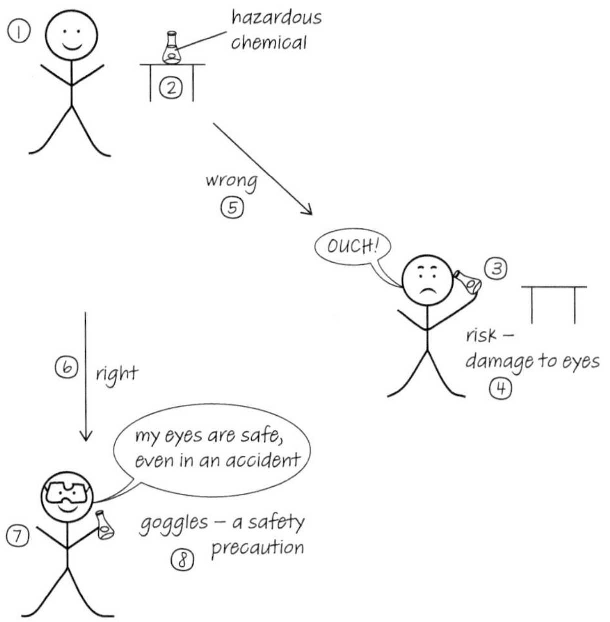

1 This is a student in the lab. They can see a chemical in a bottle they are about to use in an investigation.
2 The bottle has a red diamond symbol on it. Red diamonds are hazard symbols. This means the bottle contains a hazardous chemical.
3 This student does not know about hazardous chemicals, so picks up the bottle and shakes it. Some of the chemical gets in their eyes. This could cause permanent damage to their eyes, which are a particularly sensitive area of the body.
4 This is a risk.
5 This was the wrong way to work with a hazardous chemical that has a risk of damaging the eyes of the student.
6 There was a correct way to work.
7 The student should be wearing goggles or other eye protection in case of accidents, and should not be shaking chemical bottles unnecessarily.
8 The wearing of googles is a safety precaution, which lowers the chances of the potential risk happening.

Display the second slide, which shows the table of core questions from the Knowledge Book. Click to reveal each cell sequentially, so that the terms *hazard*, *risk* and *safety precaution* are clearly defined, and the precautions to take when working with Bunsen burners and acids are explained.

Check for understanding

Display the *CFU* slide and, using MWBs, ask questions like the ones below.

- What is a hazard?
- What is a risk?
- What risk is there when using a Bunsen burner?
- What precautions should be taken using a Bunsen burner? *There are several potential correct answers here, most of which should be used at the same time while using Bunsens. Praise all correct answers.*
- A student is making a cup of tea at home. What is the hazard?
- What risk is there in making a cup of tea?
- Strong acids are corrosive, which means they can damage your skin. What risk is there when using strong acids?
- Dilute acids (as we mostly use in school) are irritants, which means they can cause mild irritation to the skin, though more damage to eyes, which are particularly sensitive. What risk is there when using dilute acids?
- What safety precaution should be taken when using dilute or strong acids?

Independent practice review

Student Practice Book questions 94–101 (page 55).

Q	Notes
94–100	These are generic questions about risks and hazards, which help students become familiar with these terms.
101	This question encourages students to think about the effect of temperature on gas pressure from Unit C1, to consider the hazards in heating a pressurised gas container.

Core practical

Investigating the properties of elements and compounds

This practical is complex and is best carried out as a demonstration under a visualiser. A method is provided on the *Practical worksheet* and *Teacher and technician notes* on Boost – including equipment list and safety notes – if you wish to allow your students to proceed independently. You may also take a *Slow Practical* approach following the guidance below.

Students should work in groups of two or three. For each group and yourself, you will need eye protection, iron filings, sulfur powder, iron sulfide, a spatula, paper clip, bar magnet, four empty blister packs, four Petri dishes with lids, lead nitrate, zinc sulfate(VI), potassium manganate(VII), hydrochloric acid and four pipettes.

We are going to investigate how the properties of the elements iron and sulfur are different from the properties of the compound iron sulfide.

Part 1: Physical properties

1 Set up a demo station at the front. Assign the equipment by getting students to collect apparatus as needed and return to their workstations. Display the first slide and demonstrate how to set up the practical one step at a time.

2 Take a sample of iron filings and show what they look like under the visualiser. They are dark grey and slightly shiny. Tell students these filings are each made from billions of iron atoms only. Iron is therefore an element.

3 Take a sample of sulfur powder and show what it looks like under the visualiser. Sulfur is a yellow solid. This sulfur is only made from sulfur atoms, and therefore sulfur is an element.

4 A few iron filings and a small amount of sulfur can be mixed under the visualiser. Students will be able to see that this is not a uniform substance. The grey iron filings and the yellow sulfur will still be visible. Ask students if this is a compound. Why is it not a compound? Because they are not chemically bonded together.

While iron and sulfur can be reacted together in class, this can create some noxious fumes. It is not necessary for students to see this to understand that the properties of each chemical are different, so iron sulfide can be taken from a bottle or prepared in advance.

5 Take a sample of iron sulfide. Show that it is darker than iron, almost black. Unlike iron it is very brittle, and the large pieces can be easily crumbled into smaller pieces.

6 Recap the physical appearance of iron, sulfur and iron sulfide. Display the second slide and click to reveal the physical appearance part of the results table.

7 Prepare a weak magnet by rubbing a bar magnet several times in the same direction over an unbent paperclip.

A weak magnet is required because iron sulfide can display weak paramagnetic properties, especially if it was prepared in advance and some unreacted iron remains, this is why pure iron sulfide is preferable. Prepare your weak magnet shortly in advance of the demonstration so as not to add extraneous information and potential confusion.

8 Use the weak magnet to pick up some iron filings. This is because iron is magnetic.
9 Wipe the iron filings from the paperclip, and place the weak magnet into the sulfur. No sulfur is picked up by the magnet because sulfur is not magnetic.
10 Show that the weak magnet can pick the iron filings from the mixture of iron and sulfur. This is a way to separate the iron and sulfur. This again shows that they were not a compound because they were not chemically bonded to each other and could be separated.
11 Show that the weak magnet cannot pick up any iron sulfide. This is because iron sulfide is not magnetic. It does not separate the iron because in the compound the iron atoms are bonded to the sulfur atoms.
12 Recap the magnetic properties of iron, sulfur and iron sulfide. Click to reveal the the magnetic part of the results table on the slide.

Part 2: Chemical properties

The next demonstration is best carried out in microscale, because toxic hydrogen sulfide gas is produced. Demonstrate under the visualiser so you can easily highlight the key observations.

1 Place a single empty blister pack in each of four Petri dishes.
2 Label the lids of the Petri dishes: iron, sulfur, iron + sulfur mix, and iron sulfide.
3 Add a sample of each solid to the blister in each dish.
4 Add two drops of each of the following solutions to each Petri dish to the side of the blister not inside it:
   a 0.1 M lead nitrate
   b 0.1 M zinc sulfate(VI)
   c 0.002 M potassium manganate(VII).
5 Add a few drops of 1 M hydrochloric acid (HCl) to each of the blisters and add lids to the Petri dishes to contain the gases produced, so they can react with solutions a–c.
6 Demonstrate that the acid reacts with all the solids and some bubbles of gas can be seen in all of them.
7 However, the gas produced (hydrogen sulfide) is different in the reaction with iron sulfide, and this gas reacts with solutions a–c (a turns silver, b turns white, c decolorises) in a different way from the gas produced by iron and sulfur's reactions with acid.
8 Iron sulfide therefore has a different chemical property in the way it reacts with acid compared with iron or sulfur.
9 Click to reveal the chemical properties part of the results table on the slide.
10 Ask students to complete the independent practice questions on the *Practical worksheet*.

C3 Pure and impure substances

C3.1 Diffusion

Students' learning objective: *I am learning about diffusion so I can explain what happens to areas of different concentration over time.*

Students who have studied Unit B1 will already be familiar with diffusion.

Knowledge check

After working through the *General science quiz*, display the *Knowledge check* slide. The questions are best delivered one by one on mini-whiteboards (MWBs). If students do not know the answer to any of them, you should reteach the relevant concept. Answers are not on the slides because it is best to go over them together as a class.

	Question	Answer	Notes
1	Draw a particle diagram of a gas.	A diagram showing at least three particles of roughly the same size, randomly arranged, and not touching	
2	Draw a particle diagram of a liquid.	A diagram showing at least nine particles of roughly the same size, randomly arranged, and all touching	
3	Describe the movement of the particles in a gas.	Randomly, in every direction	
4	Describe the movement of particles in a liquid.	Moving over each other, but remaining touching	You are looking to see if students know that the particles are moving in liquids
5	What is diffusion?	The movement of particles from an area of high concentration to one of low concentration	If your students have studied Unit B1 already, they should have a good knowledge of diffusion. If you test this and find it to be the case, you may be able to move through this topic quickly. If your students are yet to do Unit B1 or their retention of it is weak, ensure to use all materials from this topic.

Guided explanation

What is diffusion?

Display the diagram on the slide. Tell students that you have a can of deodorant that you are about to spray. Add dots to the diagram, as shown below, to represent that it has been sprayed. (It is not necessary to demonstrate this as all students will have experienced spraying something at some point.)

Ask students who will smell the scent first. You can ask them to show you on their fingers. Students will say number 1. Tell the class that you are going to explain why it is student number 1 who smells it first, and student number 3 who smells it last.

Draw the diagram below and say you are going to talk about what is happening to the particles. Construct the diagram sequentially, with each step accompanied by an explanation, similar to the one that follows.

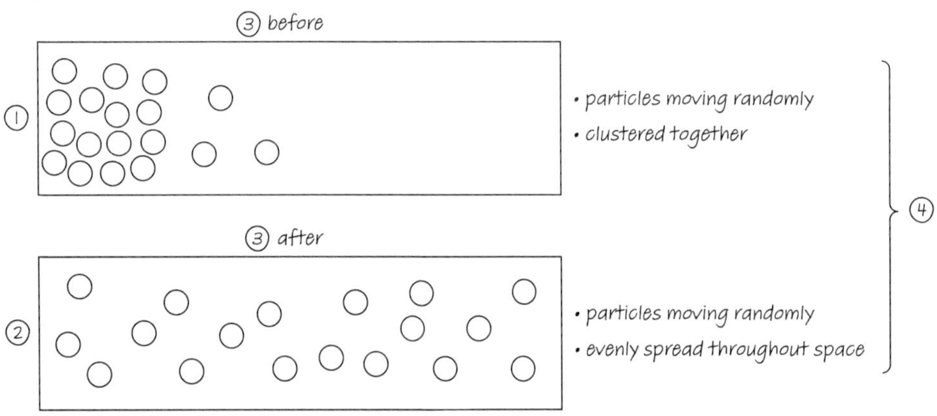

1 Say that, at first, all the particles are very close to the can, but these particles are always moving, and they are moving randomly, everywhere. You may want to add arrows showing this random movement of the particles. Explain that there are other particles in the space, but that you are not going to draw them because it makes the diagram too complicated.

2 Explain that at the beginning, when all the particles were close together in one area, we say this is an area of high concentration. Over time, they spread out into the area where there were not many particles. We call this area an area of low concentration. This happens until all the areas have the same concentration.

3 Label the *before* and *after*, so students are clear that this is the same system.

4 You may want to add labels to highlight the difference. Some students may believe that the can in some way propels the particles in a given direction. If this comes up, you can repeat the explanation, focusing on the inside of the can and the outside, explaining that the concentration inside the can is very high compared with outside. There is no mechanism inside the can to 'push' the contents out. Explain that this process happens in liquids as well.

Check for understanding

The Check for understanding (CFU) is best conducted using MWBs. Ask questions like the ones below. Tailor and adapt your questions to the class in front of you.

- What is diffusion?
- What is kinetic theory?

Display each of the *CFU* slides in turn. Ask students to identify what is wrong with each diagram, including ones in which the particles should diffuse right to left, and one where the orientation of the box is vertical instead of the ordinarily seen horizontal. For example, the issue with the first slide is that the diagram is not drawn correctly as a gas due to the regular arrangement. Students may say the particles should be spread out; emphasise that this would only be the case if diffusion has happened already.

Independent practice review

Students practise independently by answering questions 1–9 from their Practice Books (pages 56–58). Full answers can be found on the *IP answers* slides or in the answer documents on Boost. Notes are only included for questions, where useful.

Q	Notes
4b	Students will often answer questions like this by simply correcting the mistake in the original explanation, e.g. saying 'because the particles should be all spread out'. This is correct science, but does not explain why the student is wrong.

C3.2 Pure and impure

Students' learning objective: *I am learning about pure and impure substances so I can identify them. I am also learning about dissolving so I can explain how a solution is formed.*

Students will have a familiarity with these concepts from their everyday lives, but may not have formed correct definitions.

Knowledge check

After working through the *General science quiz*, display the *Knowledge check* slides and, using MWBs, deliver the questions one by one, reteaching concepts as necessary.

	Question	Answer	Notes
1	What is a compound?	Two or more different atoms chemically bonded together	
2	What is an element?	A substance made of only one type of atom	
3	What is a molecule?	A small number of atoms joined together	

4	Draw a bubble diagram for carbon dioxide.	Appropriate diagram, showing one carbon and two oxygen atoms	Provide the formula for carbon dioxide if students cannot remember it. You are not interested in whether students get the shape of the molecule correct, only that they have identified the correct number of each element, and shown that they are bonded together.
5	How many atoms are there in a molecule of carbon dioxide?	Three	
6	Oxygen exists as O_2. How many atoms are there in a molecule of oxygen?	Two	
7	How many elements are there in a molecule of oxygen?	One	
8	Define *pure*.	A substance made of only one type of particle	
9	If you put some sugar into some hot water, what happens to the sugar?	It dissolves	The concept of dissolving is a familiar concept to many students. You are not expecting students to know this answer, but if many in the class do know it, you can use the word 'dissolving' earlier in the topic without overloading students.
10	Define *mixture*.	Two or more particles not chemically bonded together	
11	Define *diffusion*.	The movement of particles from an area of high concentration to one of low concentration	
12	Does this diagram show a pure substance or an impure substance, and how can you tell? (See diagram on slide.)	Answer dependent on diagram shown	Repeat this question for each of the diagrams shown.

Guided explanation

What are pure substances?

Display the slide and click through the diagrams of the different substances sequentially, accompanying each one with an explanation, similar to the one that follows.

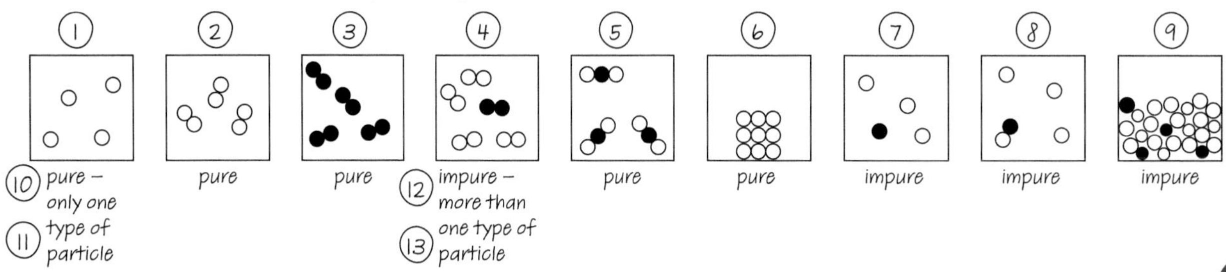

1 Explain that this gas is made of only one type of particle. Explain that it does not matter what type of particle it is, as long as they are all the same.
2 Explain that these could be individual atoms, or they could be molecules.
3 Explain that (2) and (3) are both pure, even though they are different types of particle.
4 Explain that if you mix these two together, they are no longer pure because there are different atoms, but they are not all chemically bonded, so this is not pure. Put a cross through diagram (4). Make it clear that the word 'pure' is referring to the *box (i.e. the substance) as a whole*, not any individual particle.
5 Explain that there are different atoms in carbon dioxide, but they are bonded together to form molecules. Show that the only molecules in the box are carbon dioxide molecules, and therefore this is pure.
6 Explain that it is not just gases that can be pure – it also applies to solids and liquids.
7 Explain that this substance is not pure because it is not made of all the same type of particle. Explain that we call things that are not pure *impure*. You may want to ask students to put their hands up to tell you how we know that it is not made of all one type of particle, because some students will believe that shading does not show a difference, and that they need to be different sizes or shapes.
8 Give students another example of an impure substance.
9 You may want to show an example of an impure liquid.
10 Return to substance (1) and label it as pure.
11 Add the descriptor of what pure means.
12 Label the first impure example as impure.
13 Add the definition as before.
14 Label the remaining diagrams by asking students if they think each is pure or impure.

Check for understanding

Using MWBs, ask questions like the ones below.

- Define *pure. Look out for students writing 'only one type of particle'.*
- Display the *CFU* slide and click through the diagrams of pure or impure substances. Is this substance pure or impure? Why?

Independent practice review

Student Practice Book questions 10–12 (pages 58–59).

Q	Notes
11e	Student answers should not say what the correct answer is, but instead should explain why the given answer is wrong.

Guided explanation

What is dissolving?

This explanation uses a demonstration concurrently with a diagrammatic representation at the particle level. You will need eye protection, beaker of water, sugar cube and glass rod or similar to stir with. To speed up the process, you may want to use warm, but not steaming-hot, water.

Display the slide and add to the following diagram, accompanying each step with an explanation, similar to the one that follows.

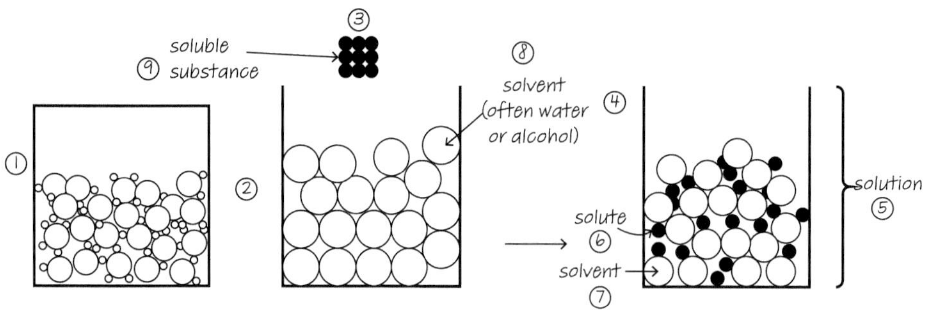

1 Display the left-hand part of the diagram. Show students a beaker half-filled with water and the diagram showing water molecules. Say that if we could magnify the water enough, this is what it would look like.

2 Explain that it takes a long time to draw all the molecules properly, so instead we show each molecule as a single circle. Erase part (1) and draw part (2).

3 Add a sugar cube to the water and stir until dissolved. Draw the particle diagram for the sugar, but do not label it yet.

4 Describe to the students what happens when the sugar goes into the water. Say that it looks like the sugar disappears, but that it does not. Draw part (4) and the arrow between the two beakers. Explain that what happens is the particles of the sugar get pulled away from each other and become held between the water molecules. We say that it has *dissolved.* (This is a simplification, but acceptable at this level. We want students to understand that the sugar has not disappeared – you just cannot see it.)

5 Explain that when we have one substance dissolved in another, we call this a *solution.* Label the solution, being clear to show that you are labelling the whole mixture, not just part of it.

6 Tell the class that a solution has two parts – the part that was dissolved and the part that did the dissolving. Remind students of the example of sugar in water. Explain that the sugar dissolved, and we call it the *solute.* Label this on the diagram.

7 Remind the class that the sugar dissolved in water, and we call the substance that does the dissolving the *solvent.* Label it on the diagram.

8 Return to part (2) of the diagram and explain that we can call it the solvent, even before it is part of a solution. Label it. Do not label the sugar (part 9) yet.

Check for understanding

Display the *CFU* slide and, using MWBs, ask questions like the ones below.

- What is part A called? *(Repeat for B–E.)*
- What happens to A when it dissolves?
- Salt is dissolved in water. Which is the solvent? Which is the solute?
- Nail polish dissolves in acetone. Which is the solvent? Which is the solute? *Repeat similar questions as necessary, using unfamiliar solvents and solutes to check if students understand the concept or if they are relying on their familiar experiences.*
- Is a solution pure or impure? Why?

Independent practice review

Student Practice Book questions 13–17 (pages 59–60).

Q	Notes
16	These questions link back to the first topic. Lower attaining groups may require you to ask some similar questions at the end of the CFU.

Guided explanation

Can all solids dissolve?

This explanation uses a demonstration concurrently with a diagrammatic representation at the particle level. You will need eye protection, beaker of water, some sand and glass rod or similar to stir with. If you used warm water in the previous demonstration, do the same here.

1. Display the slide, which shows a copy of the diagram from the previous guided explanation. Explain that sugar does dissolve in water, so we say that it is *soluble in water.* Label this on the diagram.
2. Half fill a beaker with water. Explain that some substances cannot dissolve in water – e.g. sand. Add some sand to the beaker. We can tell they cannot dissolve because we can still see them. Say that these are *insoluble in water.*
3. Explain that different substances are soluble in different solvents. Use the familiar example of nail polish. Say something like: *Nail polish is insoluble in water (which is why you can wash your hands when you are wearing it), but it is soluble in a different solvent, like nail polish remover.*
4. Point out that it is important to name the solvent you are talking about when you say a substance is soluble or not.

Check for understanding

Display the *CFU* slide and, using MWBs, ask questions like the ones below.

- A student adds an unknown substance to some water. It sinks to the bottom and stays there, even after the beaker is stirred. Is it soluble in water or insoluble in water? How can you tell?
- Repeat with named, unfamiliar substances, such as the following:
  - Copper oxide is added to some water. It sinks to the bottom and stays there, even after the beaker is stirred. Is copper oxide soluble or insoluble in water? How can you tell?
  - Sodium hydroxide pellets are added to water and the water is stirred. Eventually the pellets look like they have disappeared. Is sodium hydroxide soluble or insoluble in water? How can you tell?

Independent practice review

Student Practice Book questions 18–20 (pages 60–61).

Q	Notes
19a	You are looking for students to identify the error, not make the correction – i.e. 'they have not mentioned which solvent nail polish is insoluble in'.
19b	Here students should give the full correct answer, not just the corrected section.

Guided explanation

How do we compare how soluble substances are?

Explain that some substances are more soluble than others – in the same volume of solvent, more of the more soluble solute will dissolve. Draw the diagram below sequentially, accompanying each step with an explanation, similar to the one that follows.

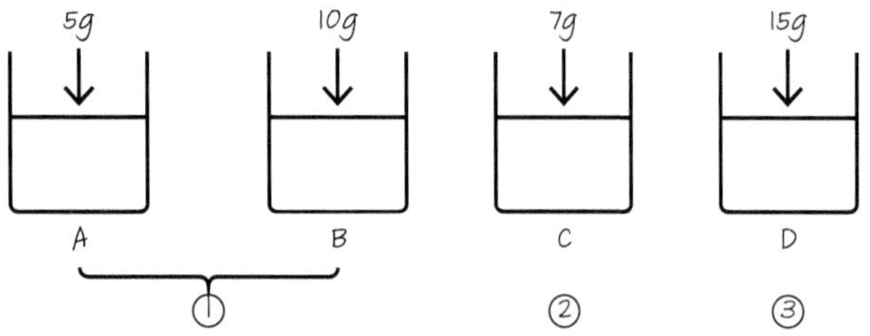

1 If we have two identical beakers, containing the same volume of water, and we can dissolve 5 g of solute A but we can dissolve 10 g of solute B, we say that B is more soluble.
2 Solute C is more soluble than A, because more can dissolve in the same volume of water, but is less soluble than B.
3 Solute D is the most soluble of these four examples because the largest mass can dissolve in the same volume.
4 We call this difference in the amount that can dissolve *solubility*.

Check for understanding

Display the *CFU* slide and, using MWBs, ask questions like the ones below.

- What is solubility?
- Which substance is the most soluble? How can you tell?
- Which substance is the least soluble? How can you tell?

Independent practice review

Student Practice Book questions 21–23 (pages 61–63).

Q	Notes
22d	This is an application question relating to something that has not been explicitly taught.

Guided explanation

What are mixtures?

Display the left-hand diagram on the slide, accompanied by an explanation similar to the one that follows, adding the labels as shown below.

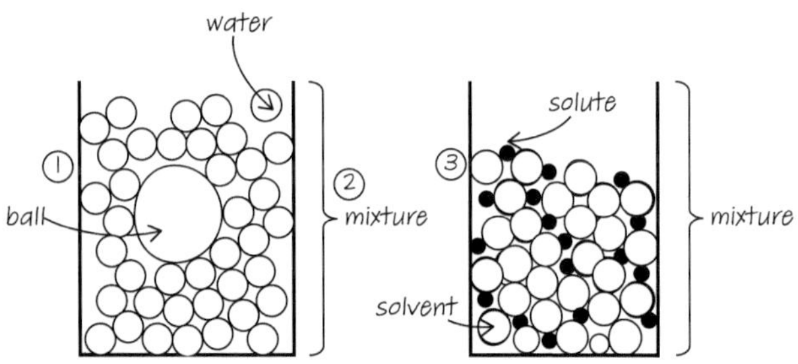

1 Say that this is the familiar example of a ball in water, and add the labels. Give students time to look at the diagram.
2 Call their attention back to you and explain that even though the ball is in the water, there are no bonds between the ball and the water and so it is easy to separate the ball from the water, e.g. with a net or a sieve. We call this a *mixture*.
3 Explain that the same applies to solutions. Click to show the right-hand diagram. Explain that the particles of the solute and of the solvent are not bonded together, so they can be separated easily, adding labels as shown.. A new substance has not been created – the solute and the solvent form a mixture.

Check for understanding

Display the *CFU* slide and, using MWBs, ask questions like the ones below.

- Which diagram(s) shows a mixture? *Show students diagrams of solids, liquids and gases to check there is no misconception that only liquids can be mixtures. Look out for students identifying D as a mixture.*
- Why is D not a mixture? *Look out for students saying, 'because it's a molecule' or 'because it's a compound'. This might be technically correct, but you want to check that students can explain why a molecule is not a mixture.*
- What type of mixture is B?
- Which one or ones are pure?
- Which are impure?

Independent practice review

Student Practice Book questions 24–26 (pages 63–64).

Q	Notes
26	For some lower-attaining groups, you may want to read aloud the term 'calcium oxalate'.

C3.3 Separation

Students' learning objective: *I am learning about different separation techniques so I can explain how to separate different mixtures.*

In this topic, we are going to see that mixtures can be separated, but different types of mixture have to be separated in different ways. The explanation is broken down into sections, one for each technique.

Knowledge check

After working through the *General science quiz*, display the *Knowledge check* slide and, using MWBs, deliver the questions one by one, reteaching concepts as necessary.

	Question	Answer	Notes
1	What is a solution?		
2	What are the two parts of a solution?	Solvent and solute	Accept either order.
3	What does *soluble* mean?	Will dissolve in a given solvent	Check that students make reference to the solvent.
4	Define *insoluble*.	Does not dissolve in a given solvent	'Does not dissolve' is insufficient.
5	What is a solvent?	Something that can dissolve another substance	

6	What is a mixture?	Two or more substances not chemically bonded	It is not necessary at this level to specify the covalent nature of the bond – just that they are bonded.
7	Some salt is added to water and stirred. **a** What is the solute? **b** What is the solvent? **c** What is the solution?	**a** The salt **b** The water **c** The salt water	
8	Describe what happens to the particles of a solute when it dissolves.	They move apart and fit between the particles of the solvent	
9	How could you tell if something was insoluble in water?	Put some of it into some water and see if it dissolves	
10	What is solubility?	How easily something dissolves	
11	Is pencil lead soluble in water?	No	Students have not explicitly learned this yet – these questions are exploratory.
12	Is ink soluble in water?	Sometimes	You can point to the fact that some inks wash off your skin easily, because they are soluble in water, but others take ages because they are either insoluble, or have low solubility.
13	Why can filtration not be used to separate a solution?	Because the particles of the solute and the solvent will both fit through the holes in the filter paper	
14	What is evaporation?	A change of state from liquid to gas, below the boiling point	Answers must include 'liquid to gas' and 'below the boiling point', but other phrasing is acceptable.
15	What happens to the state of a substance when it boils?	It turns from a liquid to a gas	
16	Define *boiling point*.	The temperature at which a substance boils	Look out for students who say 'point' instead of 'temperature'
17	What is the boiling point of water?	100°C	
18	Ethanol has a boiling point of 78°C and a melting point of -114°C. **a** What state is it at 25°C? **b** What state is it at 80°C? **c** What state is it at 100°C?	**a** Liquid **b** Gas **c** Gas	
19	Acetone has a melting point of -95°C and a boiling point of 56°C. **a** What state is it at 25°C? **b** What state is it at -96°C? **c** What state is it at 78°C?	**a** Liquid **b** Solid **c** Gas	

Filtration – How do we separate a liquid from an insoluble solid?

This explanation utilises a demonstration concurrently with a diagrammatic representation. You will need eye protection, funnel, some filter paper, beaker, some sand and water.

Remind students of the example from the previous topic of the ball in the water (you can show them a physical example of this to help too). Ask them how they would separate the ball from the water. Then show them a beaker of water and sand. Ask why they would not be able to separate the sand and water in the same way that they separated the ball and the water.

Show students a funnel and a fluted filter paper. Then draw the diagram below sequentially, accompanying each step with an explanation, similar to the one that follows.

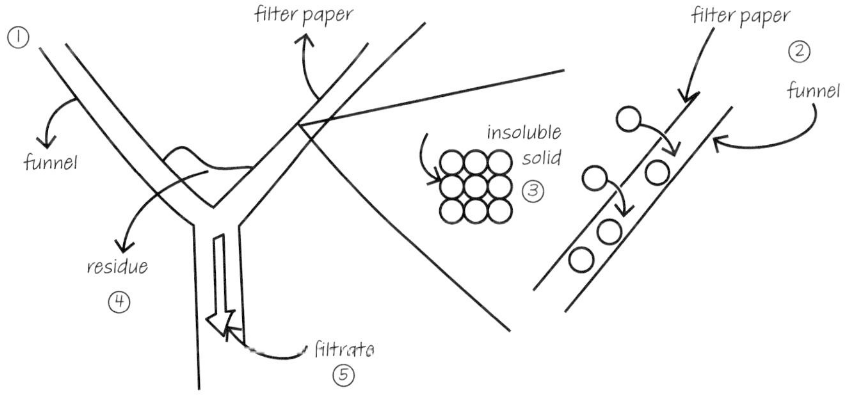

1 Label only the funnel and the filter paper. Explain that this is the apparatus used to separate an insoluble solid from a liquid.
2 Say that if we magnify the filter paper, we see that it actually has lots of little holes. Label the filter paper and the funnel on the magnified part of the diagram.
3 Explain that larger particles will not be able to fit through the gaps and will get stuck in the filter paper.
4 Pour the mixture through the filter paper. Tell them that we call this the *residue;* then add the label to the diagram.
5 Explain that things like water can fit through the filter paper because the particles are small enough to pass through the holes. Draw the arrows showing the particles moving through the holes and highlight to students the filtrate from the demonstration. Tell students that, in solutions, both particles are small enough to fit through the holes, so filters cannot be used to separate solutions. Tell them we call this the *filtrate* and add the label.

Check for understanding

Display the *CFU* slide and, using MWBs, ask questions like the ones below.

- What is A called? *(Repeat for B–I.)*
- What does A do? *(Repeat for D–F.) Look out for students thinking the funnel does the separating.*
- Why does G not fit through the filter paper?
- What would happen if G was dissolved?
- Why does the residue get left in the filter paper? *Look out for answers that do not mention particles.*
- Why can we not use a filter to separate a solution?

Independent practice review

Student Practice Book questions 27–34 (pages 64–65).

Q	Notes
33	This question is to reinforce to students that it is not the funnel that does the separating.

Guided explanation

Evaporation – How do we separate a liquid from a soluble solid?

This explanation is done via a demonstration. You will need eye protection, beaker, salt (sodium chloride), evaporating basin, spatula, tripod, clay triangle or gauze and Bunsen burner.

1. Start by dissolving several spatulas of salt into approximately 100 cm^3 of water.
2. Ask students to tell you why filtration cannot be used to separate the salt.
3. Explain that when we have a solution we cannot use filtration to separate the solvent from the solute because all the particles will flow through the holes in the filter paper. Instead, what we do is allow the solvent to evaporate, leaving behind the solute.
4. Say that, to make the solvent evaporate faster, we can heat the solution up in a special dish used for evaporation, called an *evaporating basin*. Pour the solution into an evaporating basin and draw the following diagram.

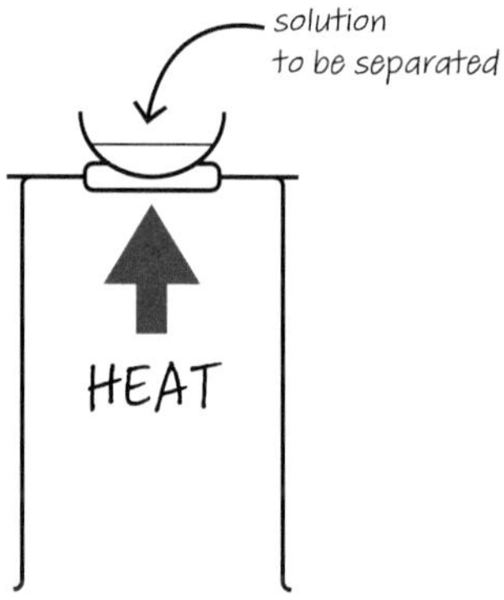

5. Begin heating the solution. Explain that it is important not to boil off all the solvent because that can damage the evaporating basin or cause little bits of the solute to escape from the basin. The solution is allowed to reach boiling and then taken off the heat to finish evaporating. The solute is left behind as crystals in the evaporating basin, so sometimes this technique is called *crystallisation*.
6. To save time, it is worth having a prepared finished sample to show students.

Check for understanding

Display the first *CFU* slide and, using MWBs, ask questions like the ones below.

- Name A. *(Repeat for B–D.)*
- What is the function of A? *(Repeat for B–D.)*
- What happens to the solute?
- What happens to the solvent?
- Salt and sugar are mixed together. Why can evaporation not be used to separate them?

Display the second slide and ask questions like the ones below.

- Which separation technique should be used?
- Why this technique?

Independent practice review

Student Practice Book questions 35–37 (page 66).

Q	Notes
36a	These questions are flexible in their response. Allow answers that are correct. You may want to mark these yourself rather than using self- or peer-assessment.
36e	Students may write a method or a description. Either is fine because you are interested in whether students can put the techniques together in a logical manner.

Guided explanation

Distillation – How do we separate two liquids?

Display the following diagram on the slide and give students time to look at it. Call their attention back to you. Add the labels sequentially, as shown below, accompanying each one with an explanation, similar to the one that follows.

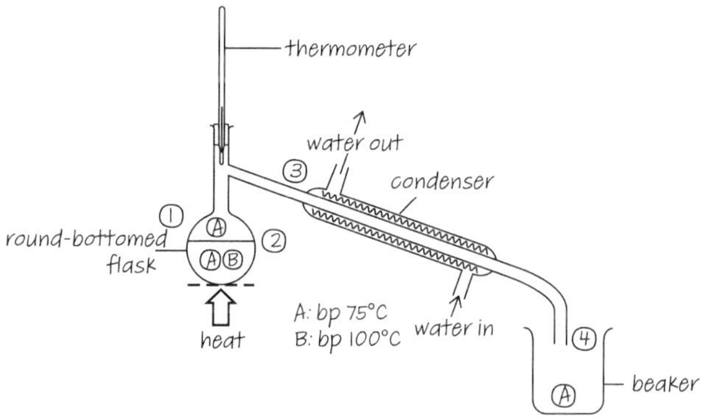

1 Explain that there are two liquids mixed together in the flask. Remind students that these could not be separated using a filter, so we will have to use a different separation technique. These liquids have different boiling points, and we can use this to separate them. Write the boiling points down so that students do not have to remember them.

2 Tell students that as the flask is heated up, substance A will reach its boiling point and turn into a gas. Show this by drawing A moving up to the neck of the flask. Say that substance B will stay as a liquid because it has not reached its boiling point.

3 Label the condenser. Explain that we need to cool substance A back into a liquid because otherwise it would be lost to the surroundings.

4 Explain that the outside of the condenser is filled with flowing water, which helps cool the gas back into a liquid, which then drips into the beaker to be collected.

Explain that you are now going to demonstrate this separating technique with two liquids.

Guided explanation

Distillation – How do we separate two liquids? *continued*

Demonstrating distillation

You will need eye protection, round-bottomed flask, quick-fix condenser, thermometer, anti-bumping granules, beaker and a mixture of water-soluble dye (such as food colouring) and water. You will also need a heating mantle or Bunsen burner.

1 Mix a few drops of water-soluble dye with 100 cm^3 of water.

2 Pour this into the round-bottomed flask, along with the anti-bumping granules.

3 Set up the apparatus as shown in the diagram below and connect the condenser, with the water entering at the bottom of the condenser.

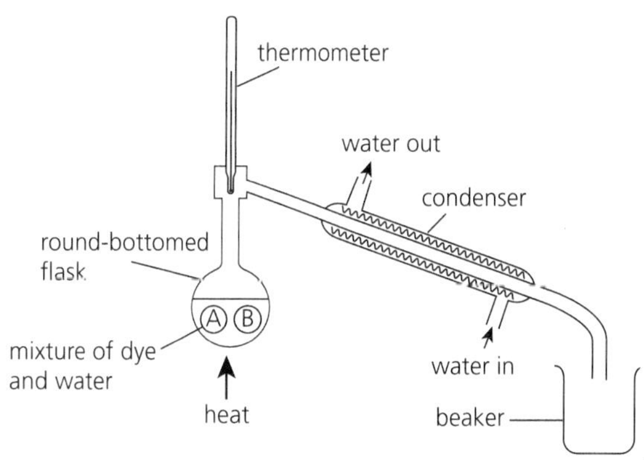

4 Begin to gently heat the round-bottomed flask. If you are using a Bunsen burner, move the Bunsen across the base of the flask to ensure an even heating and to reduce the risk of shattering the glass.

5 Ask the students what is happening to the particles as you heat the round-bottomed flask.

6 As the temperature increases, condensation will appear in the condenser. Ask students questions like:
   - What is this piece of equipment called?
   - Where did this liquid come from?
   - What is this liquid?

7 Continue heating the flask until water begins to drip into the beaker. Ask students questions like:
   - What is this liquid?
   - What happened to it in the condenser?
   - What is happening in the round-bottomed flask?
   - Why is the dye not in the beaker?
   - What does this tell us about the boiling point of the water and the dye?

   Do not heat the flask until all the water has evaporated off, because this can cause staining of the flask, depending on the ink used.

Check for understanding

Display the *CFU* slide and, using MWBs, ask questions like the ones below.

- What is distillation used for?
- What is A called? *(Repeat for B–G.)*
- What happens in A? *(Repeat for D and F.)*
- What happens to the liquids when they are heated?
- Why does only one turn into a gas?
- Why can we not use filtration to separate a mixture of two liquids?
- (Click to reveal the text.) If A has a boiling point of 58°C and B boils at 100°C:
   - Which will form a gas first?
   - Which will collect at F?
   - Why will B not be collected?
   - What temperature must we be careful not to go above?
   - Which piece of apparatus will let us know what temperature the flask is at?

Independent practice review

Student Practice Book questions 38–41 (pages 67–68).

Q	Notes
39c	The aim of this question is to get students to recognise distillation as a whole process, as opposed to just the evaporation or just the condensation 'stages'. Ensure that students use the correct terminology for the apparatus.

Guided explanation

How is chromatography used to identify pure substances?

This explanation utilises a demonstration concurrently with a diagrammatic representation. You will need chromatography paper, a pencil, ruler, coloured pens, beaker and water.

Explain that sometimes different substances have different solubilities, and that we can use these differences to separate a mixture of substances. This technique is called *chromatography*. It is worth having students repeat the word back to you, as this will get them used to using it. Tell students that this technique is often used to separate colours in ink.

1 Set up a strip of chromatography paper under a visualiser or where students can see it. It is important that we draw a start line, so we can clearly see how far each part of the mixture moved, and we need to draw it with a pencil (which is insoluble) so that it does not move up the paper with the solvent. Draw in the line on your chromatography paper.
2 Put your samples on the line and say that each sample is a mixture, and these mixtures will separate out based on their solubility, with the more soluble parts moving further up the paper.
3 Put the chromatography paper into the beaker and add a small amount of water, explaining that the water needs to be below the sample line.
4 While the separation is underway, display the slide, which shows a beaker of water and chromatography paper, as shown. Add the label 'chromatography paper'.

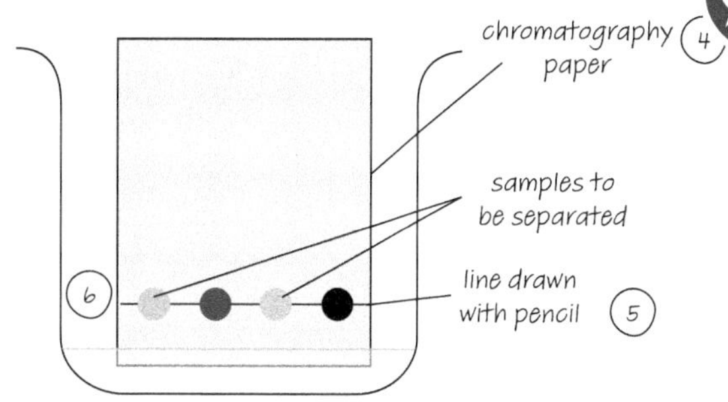

5 Indicate to students that the start line is drawn in pencil.
6 Add the 'samples to be separated'.
7 Click to reveal the completed chromatogram diagram shown below. Explain that the more soluble inks moved further up the paper than the less soluble inks and that the black ink did not move at all because it is insoluble in the chosen solvent.

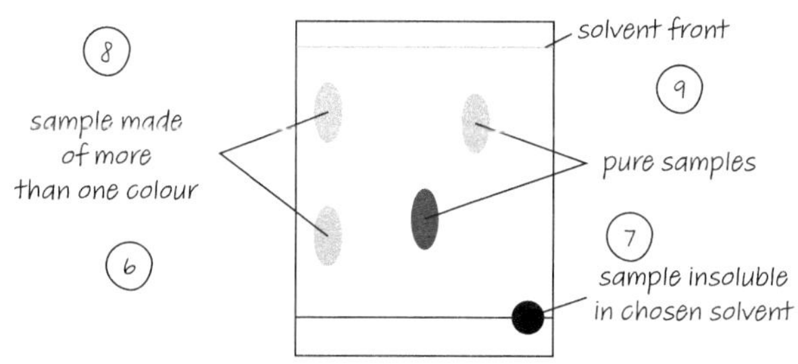

8 Point out that the green sample was a mixture of two inks – blue and yellow – which can be seen by the two colours at the end.
9 The red and the yellow were both pure because they did not separate into further colours.
10 Explain that the same colour will move up the paper the same amount, but it is not necessary at this level to mention the solvent front because R_f values will not be calculated.
11 Show students the completed physical chromatogram from the demonstration. You may want to ask students to identify which samples were pure, which were mixtures, and which were insoluble, based on the result.

Check for understanding

Using MWBs, ask questions like the ones below.

- Why is the start line for chromatography drawn in pencil?
- How can chromatography be used to identify if a substance is pure?

- Why might a sample not move up the paper during chromatography?
- Display the chromatogram on the *CFU* slide and ask questions like:
  - How many colours are in sample B? How can you tell?
  - How can you tell sample A is insoluble?
  - Which samples contain the same colours? How can you tell?

Independent practice review

Student Practice Book questions 42–47 (pages 68–69).

Q	Notes
42	Look out for students who just say, 'because it is insoluble'. They should link this statement to why it is important that the pencil line is insoluble.
46	It is important that students give reasons for their answers, to show their understanding.

Core practical

Investigating chromatography

A method is provided on the *Practical worksheet* and *Teacher and technician notes* on Boost – including equipment list and safety notes – if you wish to allow your students to proceed independently. However, for most students it will ordinarily be better to take a *Slow Practical* approach following the guidance below.

Students should work in groups of two or three. For each group and yourself, you will need eye protection, a small beaker, strip of filter or chromatography paper, sharp pencil, ruler, wooden splint, paper clip, some absorbent roll, selection of coloured pens and small amount of water.

1. Set up a demo station at the front. Assign the equipment by getting students to collect apparatus as needed and return to their workstations. Demonstrate how to set up the practical one step at a time.
2. Use the ruler to draw a pencil line approximately 1 cm from the bottom of the chromatography paper. You should check this step carefully because students will often draw a vertical line instead of a horizontal line.
3. Carefully dot three different samples along the line.
4. Pour water into a small beaker to approximately 5 mm depth.
5. Use a paper clip to attach the chromatography paper to a splint or a pencil to prevent it sinking into the water as it gets wet and soft. Check that the paper is held at the right level, by holding the paper in line with the beaker.
6. Place the chromatography paper into the water. Make sure students have not submerged their samples in the water.
7. There is a period during this practical where students simply wait for the solvent to move up the paper – you may want students to complete their independent practice in this time, or you can circulate and talk with students about their observations.
8. Once the solvent front reaches close to the top of the paper, remove it from the beaker and leave to dry on some absorbent roll. When it is dry, students should attach their paper to the *Practical worksheet*.
9. Show the students the different colours. You may want to ask them to identify features of the chromatogram using their MWBs – e.g. which sample is pure?
10. Ask students to complete the independent practice questions on the *Practical worksheet*.

P1 Energy

A few topics from the National Curriculum are covered in the additional resources on Boost. These include domestic fuel bills, fuel use and costs, heating and thermal equilibrium, energy transfer through contact/conduction and radiation, and the use of insulators. These omissions from the main text have been made to create a more logical flow with fewer digressions.

P1.1 Fuels and energy stores

Students' learning objective: *I am learning about fuels so I can understand different ways to heat things.*

Energy is a difficult topic to teach because 'energy' has colloquial and scientific meanings. We want students to get a good understanding that energy is a language we use to understand situations and a method of accounting. We will use the Institute of Physics language around energy stores and transfers, but at this point will not introduce all eight stores (nuclear, magnetic and electrostatic energy stores will be left out). We will also concentrate our efforts on simple systems by making the approximation of everything being 100% efficient. With this in mind, any additional examples need to be carefully chosen to ensure that they highlight just the stores taught. Similarly, numbers chosen for amounts of energy are not realistic to begin with, so students can concentrate on understanding concepts.

Note also that this topic is longer than most other topics in the course.

Knowledge check

As this is the first physics topic, there is no *General science quiz* at the start. Display the *Knowledge check* slide. The questions are best delivered one by one on mini-whiteboards (MWBs). If students do not know the answer to any of them, you should reteach the relevant concept. Answers are not on the slides because it is best to go over them together as a class.

	Question	Answer	Notes
1	Why do humans eat?	To gain energy	They might not use the language around energy, but that is fine for now.
2	What happens when we burn wood?	The wood changes/turns to ash	Students should avoid 'disappears'.
3	Why does this happen?	Energy is released, and smoke is given off	Students should avoid 'disappears'.
4	What do we use burning wood for?	Heating things up	

Guided explanation

What is a fuel?

Introduce the topic by asking students what a fuel is. You will need a match for the demonstration part of this explanation.

1 Show students a match and ask what will happen to the wood, and what they will see and feel, when the match is lit.
2 Tell students that in science it is not enough to say *it gets hot*. We need to explain what is happening and we use the word *energy* for this.
3 The wood in the match stores lots of energy. When we burn it, the energy is released from the wood and heats the air (or anything) that is around it. Demonstrate this by lighting the match.
4 Explain that when the energy is stored by the wood, we say it is stored in the wood's chemical energy store. When the wood is burned, this store of energy decreases.
5 Display the first slide with the animation of a match burning, and the representation of the chemical energy store decreasing.
6 Display the second slide with the animation of a phone. Explain that batteries are also called *chemical energy stores* by scientists. We increase the store by charging the phone and decrease it by using the phone.
7 We say that a fuel is any substance that releases the stored energy when it is burned or used.

Check for understanding

The Check for understanding (CFU) is best conducted using MWBs. Display the first *CFU* slide and ask questions like the ones below. Tailor and adapt your questions to the class in front of you.

- What happens to the chemical energy store if we increase the size of the candle? *Ensure that students understand that it is the wax that is the fuel, and not the wick.*
- How do you know?
- What happens to the chemical energy store if we cut the bottom half off the candle?
- How do you know?
- What happens to the chemical energy store if we burn the candle?
- How do you know?
- How could we add more energy to the chemical energy store of the candle?

Display the second slide and ask students the following questions.

- Rank the candles from largest amount of energy in its chemical energy store to smallest amount of energy in its chemical energy store.
- Justify your positioning of A. *It is not about calculating the area; it is about qualitatively checking that they are thinking about the total amount of wax that A has compared with B and C.*

There is no independent practice at this stage. If you think your class needs some consolidation, ask further questions of a similar style to the CFU.

Guided explanation

How do we measure energy?

Tell students that objects gain or lose energy from their chemical energy store, and sometimes we need to know how much this change is.

1 Explain that saying *a lot* or *a little* is not scientific, so we talk about the number of *joules* that have been gained or lost. Joules are the unit for measuring energy, like metres are the unit of distance. We give this unit a capital J.

2 Draw the previous chemical energy store example of a match and add some (simple) numbers, as shown below: 8 J when full, 0 J when empty. (You could draw extra diagrams to show intervals between 0 J and 8 J if necessary, with 4 J halfway between the two.)

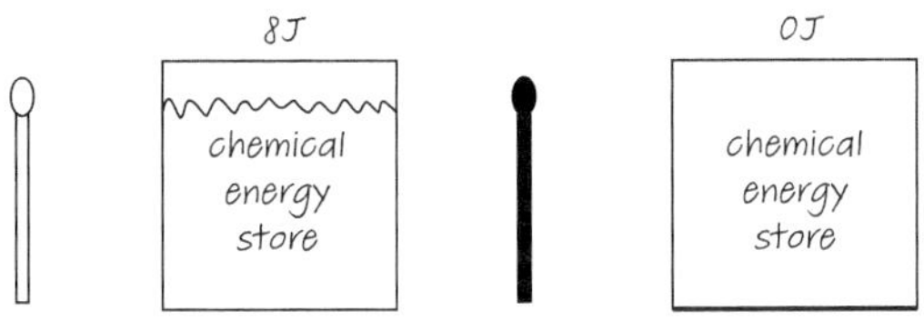

3 Explain that when we have a larger number of joules, there is more energy in the chemical energy store. And when there is more energy in the chemical energy store, we can do more with it, e.g. heat more water, send more texts, drive further.

Check for understanding

Display the *CFU* slide and, using MWBs, ask questions like the ones below.

Explain the scenario to students: a person fills up their car with petrol. It starts empty and they add petrol into it. Give students time to look at the diagrams before asking questions based on them.

- Why has the chemical energy store increased? *Answers need to be more specific than 'it's more full'.*
- How much energy would there be in the chemical energy store of the fuel tank if it were half full? *The numbers are not remotely realistic, but this helps to simplify the calculation.*
- How much energy would there be in the chemical energy store of the fuel tank if it were a quarter full?
- How much energy would there be in the chemical energy store of the fuel tank if it were three quarters full?
- Why can there not be 500 J of energy in this petrol tank?

Independent practice review

Students practise independently by answering questions 1–9 from their Practice Books (page 70). Full answers can be found on the *IP answers* slides or in the answer documents on Boost. Notes are only included for questions, where useful.

Q	Notes
5	Students may confuse how much is left with how much has been transferred out. Encourage them to draw diagrams.
9	This is a subtle concept, and may not be appropriate for all classes. The more energy we have, the more we can do. But energy is not the cause, it is a measure of what has changed.

What happens when we burn a fuel?

This explanation is done via a demonstration. You will need a match, eye protection, two beakers of water, a thermometer and a candle or spirit burner.

1 Start by taking the match and lighting it. Check that students are happy with the chemical energy store of the match decreasing. Ask where the energy that is lost from the chemical energy store is going.
2 Say that it is transferred to the air (or whatever is around the match), making it warmer. When the air gets warmer, the amount of energy it has increases.
3 Explain that this is not a chemical energy store though. The energy store to do with temperature is called the *thermal energy store*.
4 Take two identical beakers of water. Under one place a lit match and under the other place a lit candle/spirit burner.
5 Check that students understand that the candle (or spirit burner) has a larger chemical energy store than the match.
6 Show that the temperature of the water over the candle increases more than the water above the match.
7 Explain that there was energy in the chemical energy store of the fuels beforehand and, when we burned the fuels, the energy was transferred to the thermal energy store of the water. The more energy the water received, the warmer it got.
8 Draw energy store diagrams, as shown below, to illustrate what happened in both scenarios, concentrating on just the changes in energy. Show the chemical energy stores dropping to zero and the thermal energy stores increasing by the same amount as was lost.

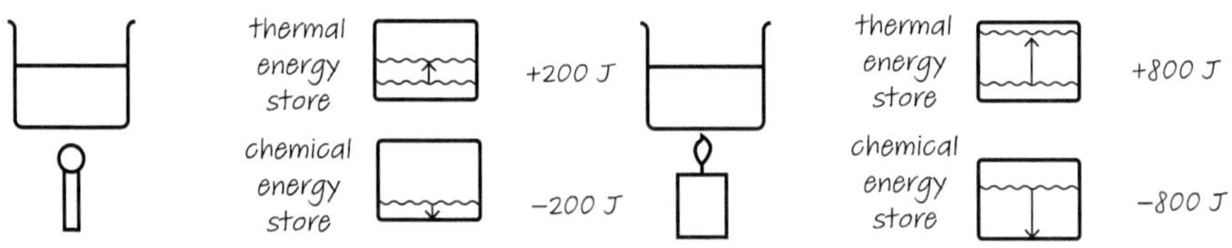

Check for understanding

Display the *CFU* slide and, using MWBs, ask questions like the ones below.

Explain the scenario to students: there are two identical blocks of metal. We want to increase the temperature of the left one by 10°C and the right one by 15°C.

- Which increase in temperature will require the bigger change to its thermal energy store?
- How do you know?
- Which will require more fuel to achieve?
- How do you know?
- If the blocks were then left to cool down, which one would lose more energy? Illustrate this with numbers if needed (e.g. the one on the left has gained 100 J and increased by 10°C, the one on the right has gained 150 J and increased by 15°C).

Independent practice review

Student Practice Book questions 10–16 (page 71).

Q	Notes
13b	We are assuming 100% efficient systems.
15	Ensure that students are giving a detailed explanation here and not just choosing A or B.

Guided explanation

How is energy transferred and how much?

Draw the diagram below to show the changes in energy stores when a candle is used to heat water, accompanied by an explanation, similar to the one that follows.

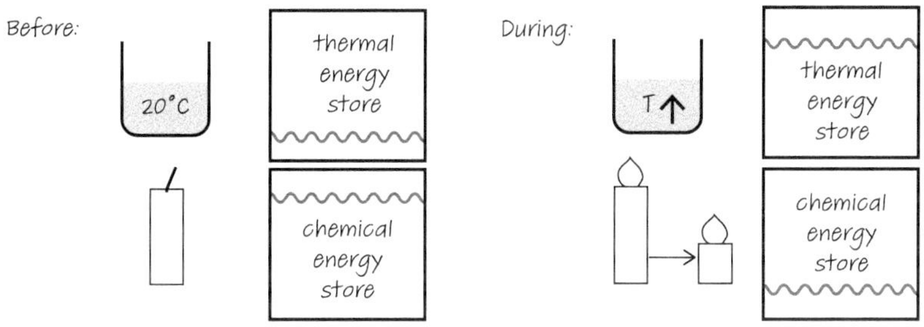

1 Point out that as the chemical energy store of the candle decreases, the thermal energy store of the water increases.
2 Explain that energy is *transferred* between the chemical energy store of the candle and the thermal energy store of the water. When we have a transfer of energy that heats up one object in this way, we call it a *transfer from hot to cold*. Emphasise that this is the direction the energy flows – from the hotter object to the colder one.

Draw the diagram below to show this chronologically, accompanied by the following explanation.

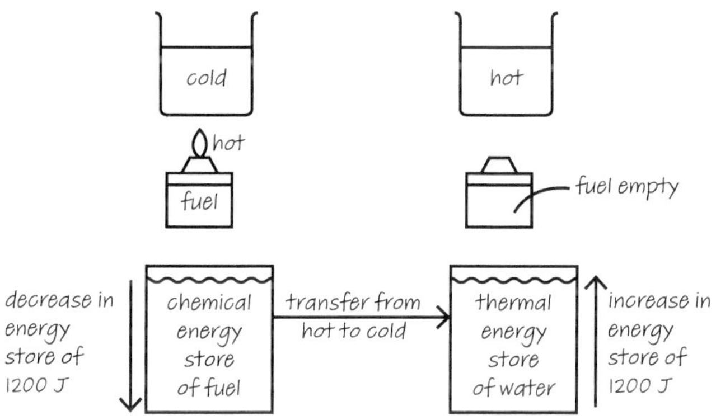

1 Link the energy stores with a labelled arrow. Say that this is the transfer from hot to cold.
2 Add numerical values to show the changes in the two stores – the chemical energy store decreases by 1200 J and the thermal energy store increases by 1200 J.
3 Explain that when energy is transferred between stores, it cannot be lost or added to along the way. We say that energy cannot be created or destroyed. The energy that started off in the chemical energy store of the candle will either still be in the chemical energy store of the candle if it has not been burned, or it will be moved into another store.
4 This means that the total amount of energy that we have will remain the same. If 100 J are lost from the chemical energy store of the fuel, the 100 J is gained by the thermal energy store of the water. This idea is called the *law of conservation of energy*.

Check for understanding

Display the *CFU* slide and, using MWBs, ask questions like the ones below.

- When the wood burns, what happens to its chemical energy store?
- What happens to the thermal energy store of the pizza?
- What is the transfer between those two stores?

- If the chemical energy store of the wood goes down by 400 J, what happens to the thermal energy store of the pizza?
- How do you know? *Students might give a lower value because some goes to the thermal energy store of the air – this may be accurate, but is more complicated than we need right now.*
- Repeat with different numbers.
- Once the pizza is out of the oven, it is at 170°C. What happens to its temperature?
- What happens to its thermal energy store?
- Where does this energy go?
- Therefore, what happens to the air?
- Which transfer is this?

Independent practice review

Student Practice Book questions 17–26 (pages 71–73).

Q	Notes
18b	Students may say 'evaporates' or 'boils'. This is not wrong, but try to steer them into thinking about smaller quantities of energy.
21c	Any idea relating to fairness, e.g. 'fair test', is fine since we have not yet introduced valid experiments.

Guided explanation

How do we measure the amount of energy in food?

Display the first slide showing pictures of water and bread.

1 Say that different foods have different amounts of energy in them. Water is important, but it does not give us the energy we need. One slice of bread will have less energy in its chemical energy store than an entire loaf. And different foods have different amounts of energy in them due to what they are made of, or how much we have – just like when we looked at candles and wood and matches earlier, all of which had different amounts of energy stored in them.

2 Say that food packaging tells us how much energy is in the food, but it does not use the unit joule to quantify it. It uses the kilojoule.

3 Display the table on the second slide as an example, or use food packaging to show this.

4 Explain that the numbers of joules we have used so far have actually been very unrealistic, because the joule is a small unit of energy.

5 One kilojoule is the same as one thousand joules, as kilo is a *prefix* that we use in front of a unit to mean 1000. So 1 kJ = 1000 J, 5 kJ = 5000 J etc.

How do we convert between J and kJ?

Write the following conversion aid on the board. (This can also be found in Topic P1.1 of the Knowledge Book.)

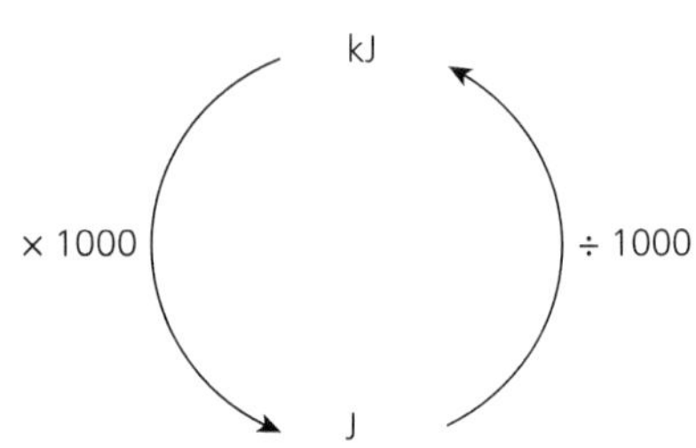

For each step below, model the conversion, e.g. converting 2 kJ to J, and ask students to do similar ones on their MWBs. Then do an equivalent 'backwards' conversion, e.g. 3000 J to kJ (where the numerical complexity is the same).

- Step 1: 2 kJ to J, 3000 J to kJ
- Step 2: 11 kJ to J, 23 000 J to kJ
- Step 3: 2.3 kJ to J, 4500 J to kJ
- Step 4: 0.4 kJ to J, 600 J to kJ

Check for understanding

Display the *CFU* slide and, using MWBs, ask questions like the ones below.

- Convert 5 kJ to J.
- Convert 0.02 kJ to J.
- Convert 3000 J to kJ.
- Convert 50 J to kJ.

Build up the numerical complexity carefully, and do lots of quick checks for understanding as you go along.

Independent practice review

Student Practice Book questions 27–36 (pages 73–74).

Q	Notes
34b	Concentrate on 100% efficient systems at this stage.

P1.2 Energy stores and transfers

Students' learning objective: *I am learning about energy stores and transfers so that I can explain how energy is involved when changes happen to objects.*

Students will now be more familiar with the concept of energy, so we will discuss further stores and transfers to move beyond fuels and heating.

Knowledge check

After working through the *General science quiz*, display the *Knowledge check* slide and, using MWBs, deliver the questions one by one, reteaching concepts as necessary.

	Question	Answer	Notes
1	What energy store changes when a fuel is burned?	The chemical energy store of the fuel	
2	What is energy measured in?	Joules, J	
3	Convert 5.2 kJ into J.	5200 J	
4	What is the law of conservation of energy?	Energy cannot be created or destroyed	Accept the *total* amount of energy stays the same (is conserved).
5	How do you make objects speed up?	Apply a force to them	Accept push/pull them or anything force related. (This question is exploratory and not expected knowledge. It will help to tailor your explanations.)
6	Why do we fall back to Earth when we jump?	Gravity pulls us back down	Specific force-related terms are not required here, just the general idea. (This question is exploratory and not expected knowledge. It will help to tailor your explanations.)

Guided explanation

What is the kinetic energy store?

Introduce the topic by showing videos of some simple moving objects, such as a rolling tennis ball, a moving car or a rocket launching. (Search for these online.)

1. Say that to get an object moving, we must transfer some energy to it.
2. Explain that when an object has this energy and is moving, we say that there is energy in the object's *kinetic energy store.*
3. A stationary (check students recall the meaning) object has zero energy in its kinetic energy store.
4. Draw some simple diagrams, such as those below, to illustrate objects moving at different speeds. You may want to use mph instead of m/s if students are not familiar with the latter. Try to ensure the objects are moving in different directions, so students can see that it has nothing to do with direction.

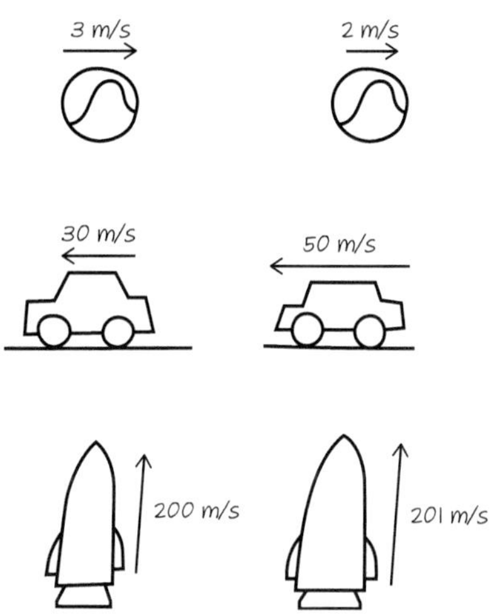

5. Explain that if an object is moving faster, it has more energy in its kinetic energy store. If the object is moving slower, it has less. It only ever has zero when it is stationary.
6. Ask students what happens to the kinetic energy store as something speeds up. It increases.
7. Ask where this energy comes from in the examples. Often it is the chemical energy store of a fuel.
8. To challenge understanding, draw a diagram to compare a stationary object with a very slow-moving object, such as the chair below.

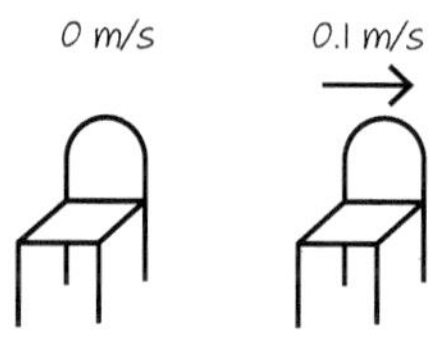

9. Explain that even the slowest objects have some, small amount of energy in their kinetic energy stores.
10. Explicitly point out that the kinetic energy store does not depend on direction, by drawing a diagram showing that two identical objects moving at identical speeds (but in opposite directions) have the same amount of energy in their kinetic energy stores.

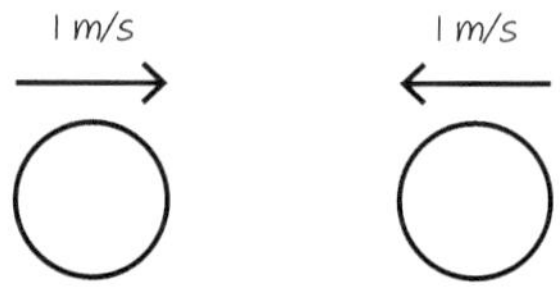

Check for understanding

Display each of the *CFU* slides in turn and, using MWBs, ask questions like the ones below.

- Which object has *more* energy in its kinetic energy store?
- How do you know?

Draw similar examples as needed to ensure students can master this.

Independent practice review

Student Practice Book questions 37–44 (pages 75–76).

Q	Notes
41	Ensure that students have *both* answers because we have not been specific about the change in question.
42	Ensure that students use the key term 'stationary' in their answers.

Guided explanation

How do we speed things up?

Remind students that when we looked at heating objects, we said that energy could be transferred between stores. Say that you will now apply the idea of transferring energy to situations where we alter speed.

1. Say that if we want an object to move faster, it needs to have more energy in its kinetic energy store.
2. So to speed something up, we need to transfer more energy to it. In this case, however, we do not do it by the transfer from hot to cold, but by a different transfer of energy.
3. Explain that to make something speed up, either from stationary to moving, or from already moving to moving faster, the object must experience a force. Tell students that we have not discussed forces yet – they will come later – but we can think of them as pushes or pulls.
4. When an object experiences an overall force, energy is transferred from the object doing the pushing/pulling to the one being pushed/pulled.
5. Draw some simple examples, such as those below, starting with objects being explicitly pushed; then move on to cars with engines, and then humans going from stationary to running. For each example, draw each energy store diagram to show:
   - the energy in the chemical energy store of the pusher or fuel decreasing
   - the energy in the kinetic energy store of the pushed object increasing
   - an arrow between the objects, labelled *transferred by a force*.

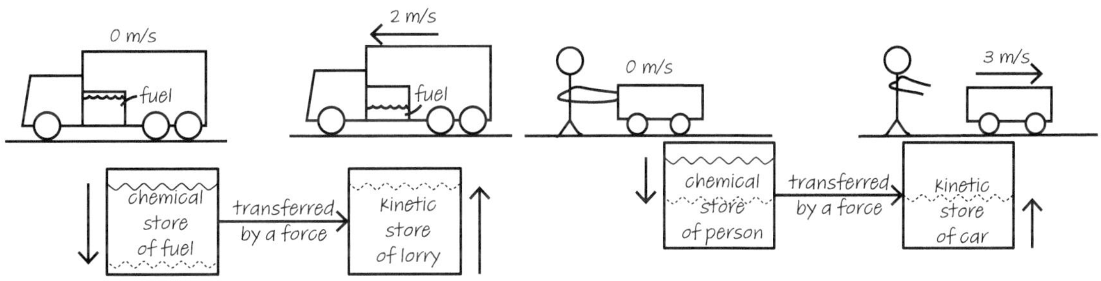

6. Explain that the transfer of energy in these examples is called *transfer by a force*, because an object is being pushed or pulled.

Check for understanding

Display the *CFU* slide and, using MWBs, ask questions like the ones below. Explain the scenario to students: a train is stationary and then starts moving and speeds up to 30 mph.

- Which energy store is decreasing? *Try to encourage students to always say 'of the', followed by the object that the store belongs to.*
- How do you know?
- Which energy store is increasing? *Try to encourage students to always say 'of the', followed by the object that the store belongs to.*
- How do you know?
- Which transfer of energy is this?

- How do you know?
- The train also warms up the surroundings. Which transfer of energy is occurring here? *This is interleaved old content. If answers to this are poor, then go back over transfers from hot to cold in this context.*

Click to reveal the energy store values. Initially there were 300 kJ in the chemical energy store. Once the train was moving, there are only 250 kJ in the chemical energy store. Ask questions like the ones below.

- How much has the chemical energy store reduced by?
- How much has the kinetic energy store increased by?

The independent practice for this section is included in the questions that follow the guided explanation below.

Guided explanation

How do we slow things down?

Start by explaining that energy in the kinetic energy store is not simply lost when something slows down, but is transferred to the thermal energy store (of the brakes and the environment). Recap by asking students what this law is called – *conservation of energy.*

1 Ask students to slow down their (moving) right hand by rubbing it over their left. They can feel that the temperature of their hands does increase.

2 Ask them questions about the stores and transfers involved, such as:
   - What store of energy does the hand have when it is moving?
   - What happens to this store of energy as it slows?
   - What did you notice happened to the temperature of your hand as you used it to slow the other hand?
   - What store of energy is involved with temperatures?
   - Did this thermal energy store of your hand increase or decrease?
   - What transfer is involved when there is a force between the objects?

3 Then add some numbers:
   - Your hand had 20 J in its kinetic energy store to begin with. It is now stationary. How much energy is in the kinetic energy store now?
   - How much extra energy must now be in the thermal energy store?

4 Draw an example, like the bike below, with an energy store diagram.

5 When you pull the brakes on a bike, the kinetic energy store decreases as you slow down. This is because of the force from the brakes. Therefore, the energy is transferred by a force and makes the brakes and wheels get hotter. This increase in temperature means it is the thermal energy store that has increased.

6 There are many good videos to supplement this explanation and to show the ubiquity of the idea, such as:
   - cars braking hard from high speeds, making their brake discs glow (search online for *'car brakes glow'*)
   - the use of the hotspot system in cricket to detect if a ball has hit the bat or pad (search online for *'cricket hotspot example'*).

Check for understanding

Display the *CFU* slide and, using MWBs, ask questions like the ones below.

Explain the scenario to students: a lorry is travelling at 25 m/s, but brakes and comes to a halt.

- Which energy store is decreasing? *Try to encourage students to always say 'of the', followed by the object that the store belongs to.*
- How do you know?
- Which energy store is increasing? *Try to encourage students to always say 'of the', followed by the object that the store belongs to.*
- How do you know?
- Which transfer of energy is this?
- How do you know?

Click to reveal the energy store values. Initially there were 100 kJ in the kinetic energy store. Ask questions like the ones below.

- How much has the kinetic energy store reduced by?
- How much has the thermal energy store increased by?
- What will have happened to the temperature of the brakes?

Independent practice review

Student Practice Book questions 45–51 (pages 76–77).

Q	Notes
48b	Remember that we are assuming 100% efficiency.

Guided explanation

What happens when we lift things up?

Start by dropping a ball from a height.

1 Ask students what happens to the speed of the ball as it falls.

2 Ask which store of energy increases as it gets faster.

3 Remind students that we know energy cannot just appear, so this gain of energy in the kinetic energy store of the ball must have come from somewhere.

4 Say that the energy came from a store called the *gravitational energy store.*

5 Explain that when you lifted the ball up, your chemical energy store decreased, and this energy was transferred to the gravitational energy store of the ball as it was lifted higher and higher. Illustrate this by drawing the following diagram.

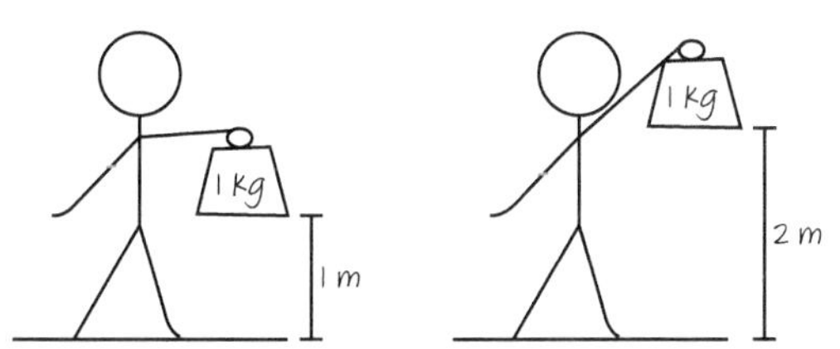

6 Say that raising an object higher and higher requires more and more energy to be transferred to it. This is because we are fighting against the gravitational pull of the Earth, which is pulling the object back down to Earth.

7 Draw a second diagram, like the one below, to split the action of lifting objects up into steps.

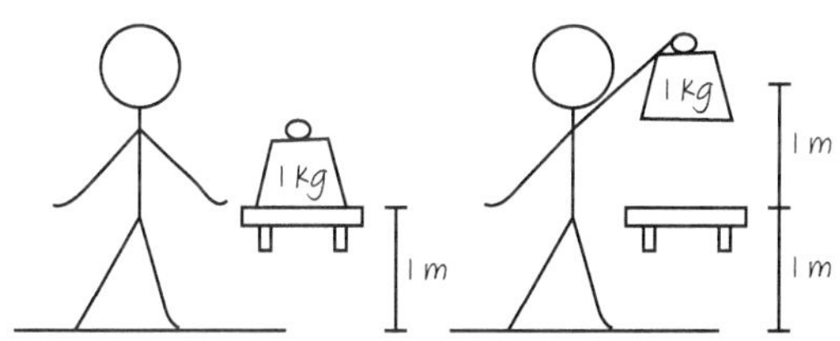

8 Say that to lift an object 2 m up, we could first lift it 1 m and place it on a shelf, before lifting it 1 m more. This clearly requires more energy to be transferred from ourselves. The gravitational energy store of the object increases the more we lift it.

Check for understanding

Display each of the *CFU* slides in turn and, using MWBs, ask questions like the ones below.

- Which object has energy in its gravitational energy store?
- How do you know?
- Which object has *more* energy in its gravitational energy store?
- How do you know?

Independent practice review

Student Practice Book questions 52–60 (pages 77–79).

Q	Notes
57a	Remember that we are assuming 100% efficiency.
57, 58	These bring together many parts of the topic covered so far.

Guided explanation

What happens if we stretch, squash or twist things?

The formal vocabulary around extensions and compressions, as well as the words *elastic* and *inelastic*, will be left until Unit P3 on forces and Hooke's law.

Eye protection should be worn. Show students a piece of elastic, a spring that works under compression and a 30 cm plastic ruler. Extend, compress and twist each of them in turn, more and more (do not take them past their elastic limit).

1. Discuss the change in your chemical energy store that must be happening, and use this to motivate the change in elastic energy store.
2. Explain that more stretch, compression or twist leads to more energy in the elastic energy store.
3. Illustrate this by drawing the simplified energy store diagram below.

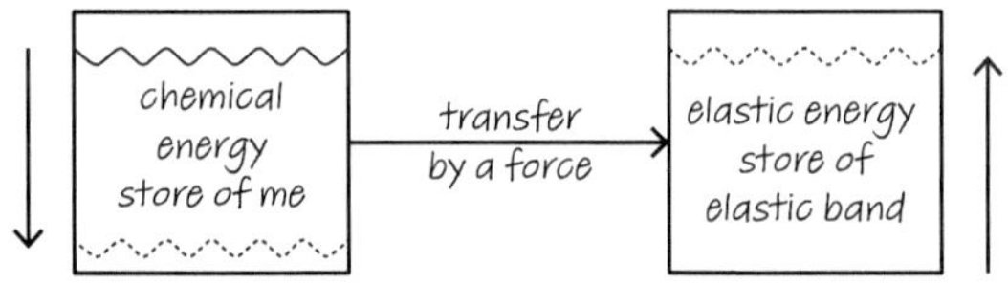

4. If you feel that more demonstrations are needed, consider other ways of stretching and compressing that use simple energy stores, such as:
   - Hanging larger and larger masses from elastic or springs, dropping them from the unstretched length to see the increase in extension with increased mass (gravitational energy store to elastic energy store via transfer by a force).
   - Dropping larger and larger masses from the same height onto a spring that works under compression (gravitational energy store to elastic energy store via transfer by a force). Take into account where the masses might fall.
   - Rolling a ball faster and faster into a piece of elastic stretched between two fixed points (kinetic energy store to elastic energy store via transfer by a force).

Check for understanding

Display each of the *CFU* slides in turn and, using MWBs, ask questions like the ones below.

- Which is more stretched or compressed out of the two options?
- Which has more energy in its elastic energy store?
- How do you know?

Independent practice review

Student Practice Book questions 61–67 (pages 79–80).

Q	Notes
61	Ensure that students give all three answers: stretch (extend), squash (compress) and twist.
62c	If students are ever confused about 'which transfer' it is, ask them: 'is there a force involved?'.
67b, d	Note the *always* in the question. The elastic and kinetic energy stores are repeatedly increasing and decreasing, and so are not either *always* increasing or *always* decreasing.

Guided explanation

How else can energy be transferred?

Students will not have a lot of the background knowledge to truly understand transfers by waves and by electric current, but they are so ubiquitous in even the simplest of examples (lightbulbs, speakers etc.) that we must mention them briefly. (This is in contrast to the nuclear, magnetic and electrostatic energy stores that we have omitted entirely from the course because they can generally be avoided.)

Do not use examples involving multiple changes. Use simple examples with a 'start point', and 'end point' and one energy transfer.

Work through the following two examples with students:

1 Draw a diagram of a battery-powered remote-controlled car starting stationary and moving at a later time.

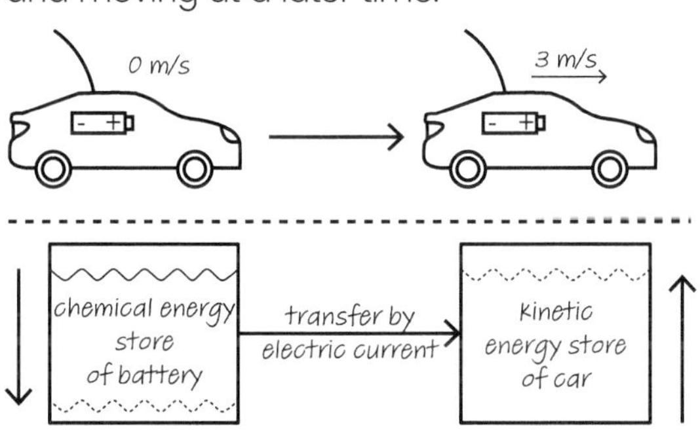

2 Add the energy store diagram with the two stores that are changing. Explain that as the car moves faster, the kinetic energy store is increasing. This is because of the battery being used, so the chemical energy store is decreasing.

3 Then link the changing stores using an arrow labelled *transfer by electric current.*

4 Draw a similar diagram for a battery-powered heater (chemical energy store of battery decreasing, thermal energy store of heater increasing).

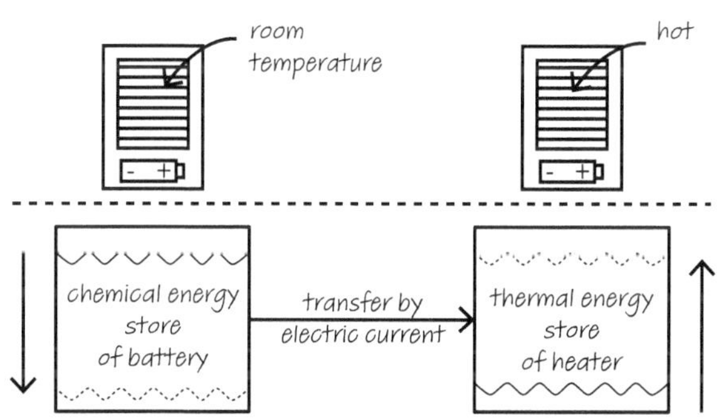

5 Ask students how the energy was transferred from the battery to the heater. *Transfer by electric current.*

6 When dealing with mains electricity, since the initial store of energy is not clear (it could be from one of many energy resources), it can be avoided entirely.

7 Draw the charging of a phone (chemical energy store of battery increases), as shown below.

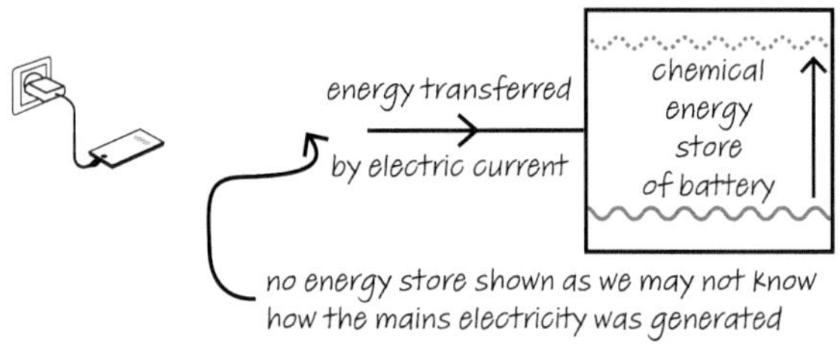

8 Ask students how the energy was transferred to the battery? *Transferred by electric current.*

9 Explain that sound and light are both examples of waves (they radiate out from an object), and so are treated identically. Depending on the class, you could discuss how transfers by waves will always (eventually) lead to an increase in the thermal energy store of the environment. Examples include:

- Battery-powered lightbulb (chemical energy store of battery decreases; energy transferred by a current to the bulb; energy transferred by waves (light)).
- Battery-powered speaker (chemical energy store of battery decreases; energy transferred by a current to the speaker; energy transferred by waves (sound)).

Check for understanding

Display the *CFU* slide and, using MWBs, ask questions like the ones below.

- When is energy transferred by an electric current?
- Which energy store is associated with a battery? *This is interleaved with old content, so go back over this if necessary.*
- Which transfer is involved when one object heats up another? *Ensure that students do not say thermal energy store, because we are looking for the transfer between stores.*
- Give two examples of energy being transferred by waves.
- If I push an object to make it move, which transfer of energy is this? *Ensure that students do not talk about the chemical energy store of me or the kinetic energy store of the object (as these are not the transfers).*
- Which energy store decreases when an object falls to the ground?

Independent practice review

Student Practice Book questions 68–77 (pages 80–82).

Q	Notes
70c	Accept only transfer by sound. Do not accept transfer by light.

P1.3 Power

Student's learning objective: *I am learning about power so that I can calculate how quickly energy can be transferred between stores.*

In our everyday lives, people often use the word 'power' and 'energy' interchangeably. However, in science they have very different meanings, which we must be careful to unpack with our students. To ensure students do not conflate the concepts of energy and power, the National Curriculum section on kW hour as a unit of energy is covered in the additional materials on Boost.

Knowledge check

After working through the *General science quiz*, display the *Knowledge check* slide and, using MWBs, deliver the questions one by one, reteaching concepts as necessary.

	Question	Answer	Notes
1	What is distance measured in?	Metres, m	Look out for 'meter' (measuring instrument) rather than 'metre' (unit of length). The letter 'm' should be lower case, not upper case. If other units are given (e.g. cm, km), praise students but point out that m is the standard unit.
2	What number does the prefix 'kilo' represent?	1000	
3	Convert 5.9 km into m.	5900 m or 5900 metres	Look out for students multiplying/dividing by 100 or dividing by 1000.
4	What is energy measured in?	Joules, J	Look out for lower case 'j'. The symbol for joules is upper case 'J' since it is named after the scientist James Prescott Joule (1818–1889).
5	Convert 3700 J into kJ.	3.7 kJ	Look out for students multiplying/dividing by 100 or multiplying by 1000.

Guided explanation

What is power?

Display the slide with the animation incorporated into the fan images. Click through the stages of the animation sequentially, accompanying each step with student input or explanation, similar to below.

1. We have a fan with two buttons marked 'High' and 'Low'.
2. When we press the 'Low' button it turns slowly.
3. When we press the 'High' button it turns quickly.
4. Ask students about the stores of energy and transfers involved.
5. The button on the fan allows us to swap between two different settings, but the stores of energy and the transfers remain the same.
6. Ask students on which setting the battery will run out the fastest. This is a very intuitive question, and they are all likely to see immediately that it is the one on the high setting.

Draw the diagram below sequentially, with each step accompanied by an explanation, similar to the one that follows.

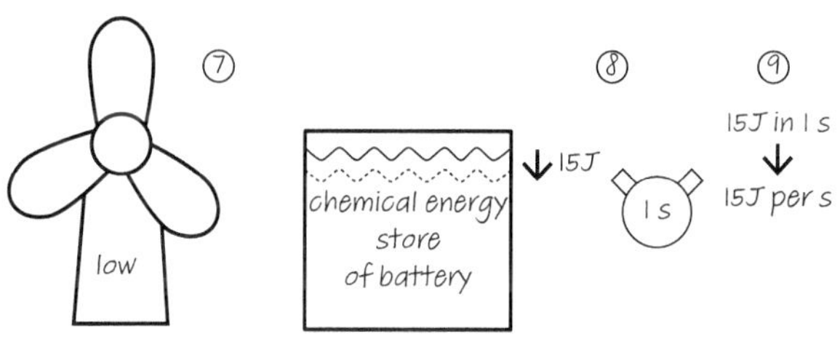

7. Starting with the fan on the low setting, we see that the chemical energy store of the battery will decrease.
8. If we measure how much it decreases for 1 s, we see that it might go down by 15 J.
9. So we say the chemical energy store has gone down by 15 J in 1 s. We can write this using more mathematical words as 15 J per s.

Explain that we say this is the *low power* of the fan, because not much energy is transferred each second.

Construct a similar diagram for the fan on high power, as shown below.

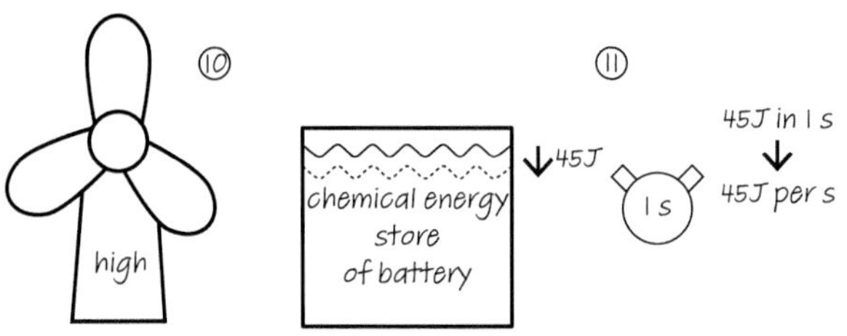

10 This time, if we measure for 1 s, we see that the chemical energy store of the battery decreases by more than last time – 45 J. More energy has been transferred in 1 s.

11 We say the chemical energy store of the battery has decreased by 45 J in 1 s. We can write this as 45 J per s.

Explain that we say this is the *high power* of the fan because *more* energy is transferred each second.

Construct a similar diagram for a microwave, as shown below (which we have now simplified to remove the energy diagram scaffold).

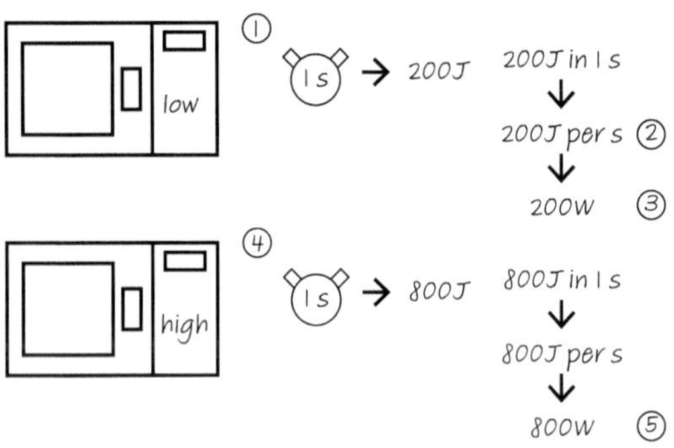

1 This microwave has a low-power setting. In 1 s the microwave will transfer 200 J.

2 We can write this as 200 J per s.

3 In science, we often name units after scientists, e.g. the units of energy, Joules, are named after James Joule. Instead of saying joules per second each time, we give power its own unit, which also commemorates the work of the scientist James Watt. We say 1 J per s is 1 W. So we write 200 J per s as 200 W.

4 The microwave also has a high-power setting. In 1 s the microwave will transfer 800 J.

5 800 J in 1 s can be written 800 J per s, and we call this 800 W.

Finally, construct a diagram like the one below, with several examples of objects with different power (not all electrical, so students do not assume that this is only about technological objects).

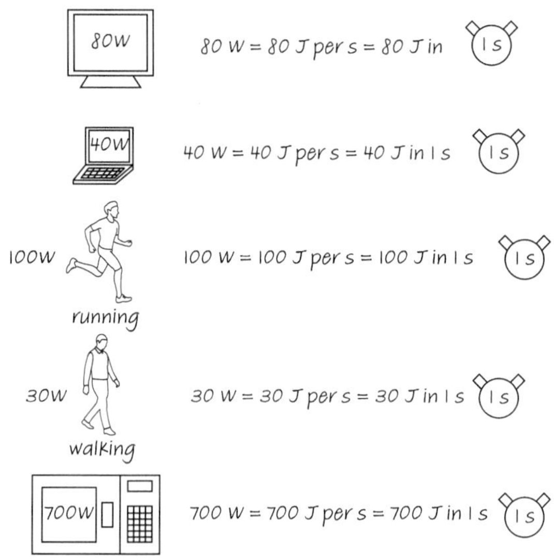

For each example we will start in the reverse direction and give the power. We will then translate this in the reverse direction to what we did earlier, e.g.:

1 The TV has a power of 80 W. We know this is the power of the TV because the units are watts (this is important to stress – it is why units are useful).

2 A power of 80 W means 80 J per s.

3 This means that in 1 s, 80 J are transferred.

Compare the power of different objects as you go, for example:

- The laptop has a lower power than the TV because it transfers less energy per second.
- The microwave has the highest power of any of these objects because it transfers the most energy per second.

Check for understanding

Display the first *CFU* slide and, using MWBs, ask questions like the ones below.

- What units do we use to measure power?
- Which toaster has a higher power?
- How do you know?
- What is the power of toaster A? What is the power of toaster B?

Display the second slide and ask questions like the ones below.

- Which fan has a higher power?
- How do you know?
- What is the power of fan A? What is the power of fan B?
- What do we mean by high power?
- What do we mean by low power?

There is no independent practice at this stage. If you think your class needs some consolidation, ask further questions of a similar style to the CFU.

Guided explanation

What is power? (*continued*)

In this section, we will begin to build more towards the power equation. We will not explicitly be using the equation, but it will put students in a position whereby they have a better understanding of power, so that the equation makes more physical sense.

Draw a diagram like the one below, with examples of different objects (try to keep the objects varied and all electrical). Construct the diagram sequentially, with each step accompanied by an explanation, similar to the one that follows.

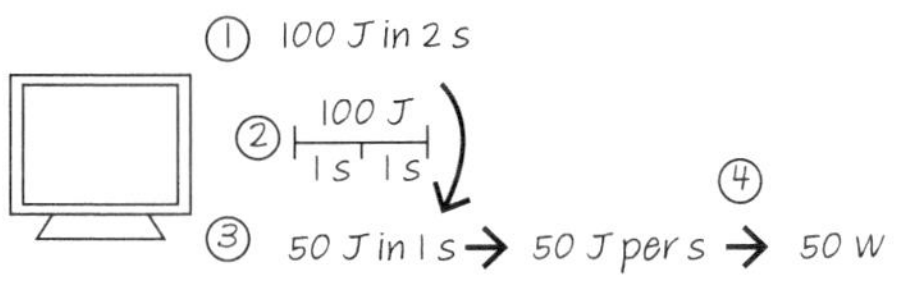

1 The TV transfers 100 J in 2 s. We cannot immediately tell what the power is because we know what it transfers in 2 s, not 1 s.

2 (Optional) Draw a simple bar model if students need support in seeing how the 100 J will be divided up per second.

3 100 J in 2 s, means there must be 50 J in 1 s.

4 Earlier, we said 50 J in 1 s means the same as 50 J per s, which is 50 W.

Repeat this process for the toaster and hairdryer.

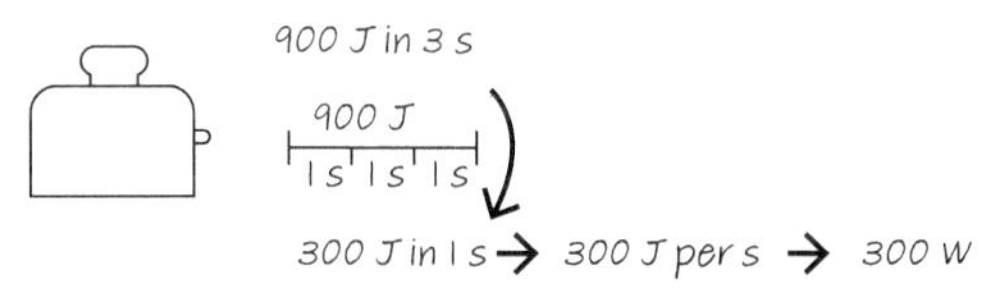

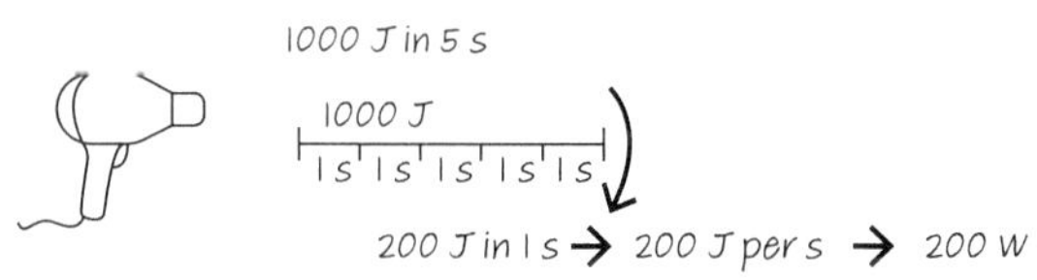

Check for understanding

Display each of the *CFU* slides in turn and, using MWBs, ask questions like the ones below.

- Which toaster has a higher power?
- How do you know? *This is a fairly straightforward comparison because B transfers more energy than A in the same time.*
- Which bulb has a higher power?

- How do you know? *This is a fairly straightforward comparison because B transfers the same energy as A in a shorter time.*
- Which TV has a higher power?
- How do you know? *Encourage students to draw out a number line like those in your examples.*

Independent practice review

Student Practice Book questions 78–91 (pages 82–83).

Q	Notes
89	Ensure that students see that only one of the values needs converting. The direct comparison can only be made once the units are the same.

Guided explanation

How do we find the power equation?

Construct a series of examples, as shown below, with each step accompanied by an explanation, similar to the one that follows.

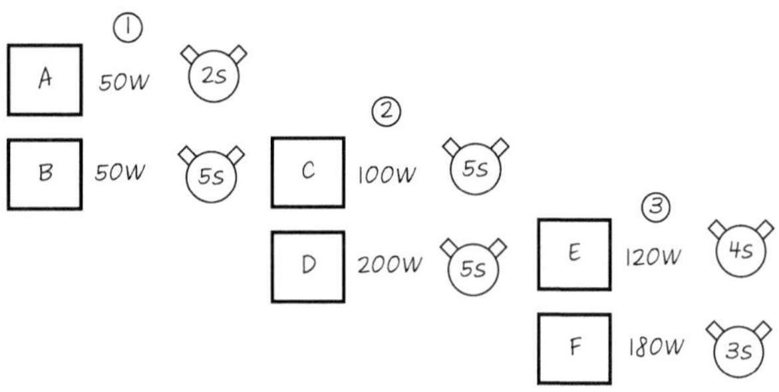

1. Object A has a power of 50 W and is left on for 2 s. Object B also has a power of 50 W but is left on for 5 s. Which transfers more energy? (Students should intuitively pick B, but guide them through 50 W means 50 J per s if necessary.)
2. Object C has a power of 100 W and is left on for 5 s. Object D has a power of 200 W and is also left on for 5 s. Which transfers more energy?
3. Object E has a power of 120 W and is left on for 4 s. Object F has a power of 180 W and is left on for 3 s. Which transfers more energy? (This one is not immediately obvious, and requires some mathematical manipulation. This is where the equation for power is useful.)

Construct the following diagram, with each step accompanied by an explanation, similar to the one that follows.

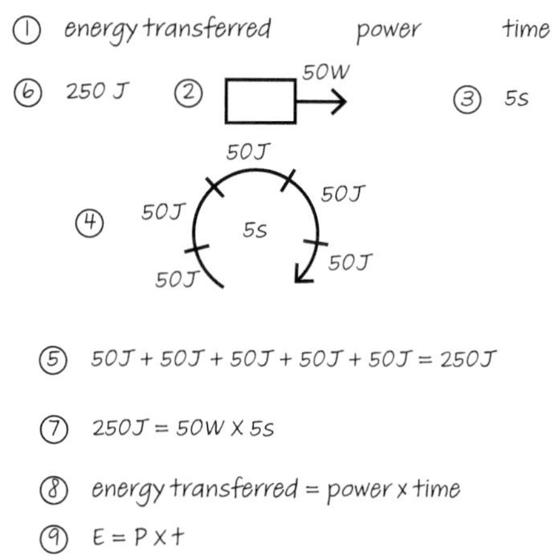

1. We are going to look at an object – its power, the time it is on for and how much energy it transfers.
2. The device has a power of 50 W.
3. We leave it on for 5 seconds. Let us look at how much energy it transfers.
4. First, we can draw a little wheel – a bit like a clock face – that shows the 5 seconds. In each of those seconds, it transfers 50 J (because its power is 50 W).
5. How much energy will it transfer in 5 seconds? To work this out, we add up all of the 50 J.
6. This equals 250 J.
7. A quicker way of doing this would be to multiply the power by the time.
8. This gives us an equation that we can use whenever we see power, time and energy transferred.
9. We can write the equation in words and in symbols.

Check for understanding

Display the first *CFU* slide and ask questions like the ones below.

Construct the model as above, but for a device with a power of 20 W that is left on for 4 s. Use verbal questioning so that students start to pick up the model, e.g.:

- How many joules does it transfer every second?
- How many sections do I split my clock face into?

Display the second slide and work through the next example, for a device with a power of 40 W that is left on for 6 s, this time using MWBs to bring in the rest of the class.

- How many joules does it transfer in one second?
- How many sections do I need on my clock face?
- Label each section of the clock face with the number of joules each second.
- How much energy is transferred in total?

Display the third slide and repeat the steps above for a device with a power of 500 W that is left on for 10 s.

Independent practice review

Student Practice Book questions 92–97 (page 83).

Q	Notes
95–97	Students should focus on 'how many per second'. We are not expecting proper calculations here, just motivating the equation.

Guided explanation

How do we use the power equation?

Complete the worked example on the slide, as shown below, to model the EVERY method and demonstrate how we use the power equation. Explain each step line-by-line.

A TV set has a power of 800 watts and is switched on for 30 seconds. Calculate the energy transferred.

Equation	energy = power × time $E = P \times t$
Values	$E = ?$ $P = 800\,\text{W}$ $t = 30\,\text{s}$
Enter values	$E = 800 \times 30$
Result	$E = 24\,000$
Y(units)	$E = 24\,000\,\text{J}$

Check for understanding

Display each of the *CFU* slides in turn and, using MWBs, ask questions like the ones below.

- Start by asking students to write EVERY down the side of their whiteboard.
- Ask them to complete each of the steps of EVERY, showing their whiteboard to you.
- Ensure you use *Show Call* to emphasise students who are showing full working.

- Suggested values for further checks if needed:
  - power: 80 W, time: 20 s
  - power: 550 W, time: 15 s
  - power: 10.5 W, time: 45 s

Do not allow students to progress to the independent practice until they are getting questions like these correct on their MWBs.

Independent practice review

Student Practice Book questions 98–107 (pages 83–84).

Q	Notes
99–107	With all calculations, ensure students are showing their working out. Show exemplary student work and critique those who need a little feedback (e.g. forgetting units or forgetting that there needs to be something on either side of the '=' sign).
105	Ensure that students do not get distracted by the 400 m, which is extraneous.
106c	Ensure that students see that the question is saying 'and then', so the amounts of energy are added together.

Guided explanation

How do we convert kilowatts to watts?

Students learned how to convert J into kJ and vice versa in Topic P1.1.

Demonstrate the conversion by writing 10 kW on the board and asking students to show you on MWBs what they think it would be in W. If they can do this with ease, try a few more conversions, such as:

- 1000 kW
- 0.5 kW
- 13 000 W (to convert into kW)
- 175 W

If they are struggling, present a full explanation, as in Topic P1.1, by drawing the diagram below.

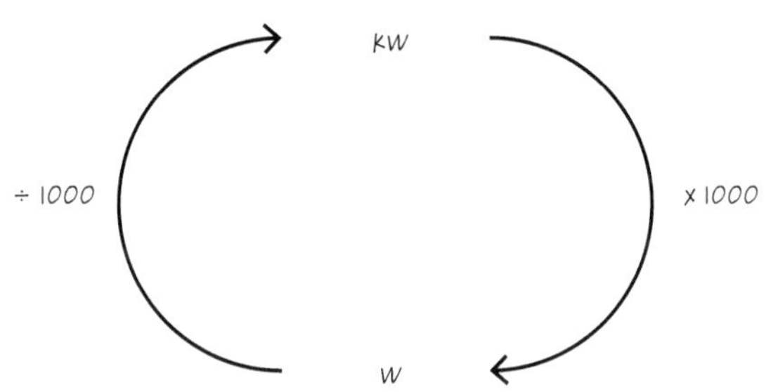

Check for understanding

Display the *CFU* slide and, using MWBs, ask questions like the ones below.

- Ask students a range of conversion values (any numbers will do). *Ensure they can convert from kW to W, and vice versa.*
- Another good way to CFU with unit conversions is to give students incorrect scenarios. For example: a student converts 2000 kW into W and gets 2 W. What mistake has the student made?

Do not allow students to progress to the independent practice until they are getting questions like these correct on their MWBs.

Independent practice review

Student Practice Book questions 108–13 (pages 84–85).

Q	Notes
108–13	Ensure students show all working out for calculations.

Guided explanation

How do we convert from W to kW in the power equation?

We will now combine knowledge of the power equation with knowledge of conversions (both kilowatts to watts and minutes to seconds) to get students to do more complex calculations.

To explain how to do kilowatt-to-watt conversions within a calculation, complete the worked example on the slide, using the EVERY method and explaining each step line-by-line.

A kettle has a power of 2.7 kilowatts and is switched on for 70 seconds. Calculate the energy transferred.

Equation	$E = P \times t$
Values	E = ? P = 2.7 kW t = 70 s Show how to multiply by 1000 to convert 2.7 kW to 2700 W
Enter values	$E = 2700 \times 70$
Result	$E = 189\,000$
Y(units)	$E = 189\,000$ J

Check for understanding

Display each of the *CFU* slides in turn and, using MWBs, ask questions like the ones below.

- For Example 1, start by verbally asking students to fill in the next steps.
- For Example 2, students should be doing the calculation alone.

Do not move on until students get the second example correct by themselves. If necessary, use more examples with any values. The independent practice for this section is included in the questions that follow the final conversion guided explanation below.

Guided explanation

How do we convert minutes to seconds?

We will now look at time conversions between minutes and seconds.

Display the example on the slide and take suggestions for how to solve it. Following suggestions, demonstrate the simple conversion by drawing the following diagram.

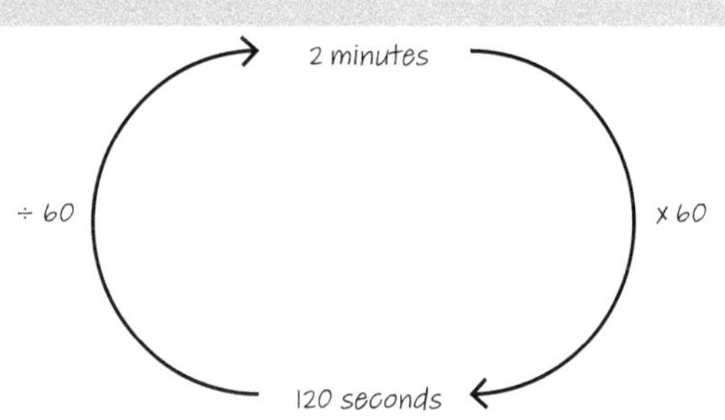

Students may already have knowledge of time conversions, but it is worth constructing a diagram like the one shown as a general tool. (This can also be found in the Working Scientifically section of the Knowledge Book.)

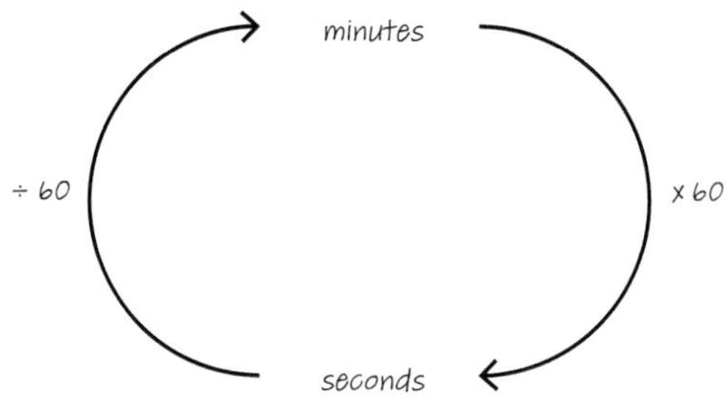

Check for understanding

Display the *CFU* slide and, using MWBs, ask questions like the ones below.

- Ask students a range of conversion values (any numbers will do). *Ensure they can convert from minutes to seconds, and vice versa.*
- Another good way to CFU with unit conversions is to give students incorrect scenarios. For example: a student converts 60 minutes into seconds, and gets 1 s. What mistake has the student made?

Do not allow students to move on until they are getting questions like these correct. The independent practice for this section is included in the questions that follow the final conversion guided explanation below.

Guided explanation

How do we convert minutes to seconds in the power equation?

Complete the worked example on the slide, using the EVERY method and explaining each step line-by-line. Depending on the class, you may want to bring them into the explanation via questioning or MWBs.

A microwave has a power of 700 watts and is switched on for 3 minutes. Calculate the energy transferred.

Equation	$E = P \times t$
Values	E = ? P = 700 W t = 3 min Show how to multiply by 60 to convert 3 min to 180 s.
Enter values	E = 700 × 180
Result	E = 126 000
Y(units)	E = 126 000 J

Check for understanding

Display each of the *CFU* slides in turn and, using MWBs, ask questions like the ones below.

- For Example 1, start by verbally asking students to fill in the next steps.
- For Example 2, students should be doing the calculation alone.
- Another good way to check for understanding with unit conversions is to give students incorrect scenarios. For example: a student converts 2000 kJ into J, and gets 2 J. What mistake has the student made?

Do not allow students to move on until they get questions like these correct. If necessary, use more examples with any values. The independent practice for this section is included in the questions that follow the explanation below.

Guided explanation

How do we convert joules to kilojoules?

By this point, students should be adept at unit conversions.

Display the example on the slide and ask students how to solve it. Work through it together and show them how to convert the units at the end.

A microwave has a power of 950 watts and is switched on for 1.5 minutes. Calculate the energy transferred. State your answer in kilojoules.

Equation	$E = P \times t$
Values	$E = ?$ $P = 950\,\text{W}$ $t = 1.5\,\text{min}$ Show how to multiply by 60 to convert 1.5 min to 90 s
Enter values	$E = 950 \times 90$
Result	$E = 85\,500$
Y(units)	$E = 85\,500\,\text{J}$ Show how to divide by 1000 to convert 85 500 J to 85.5 kJ

Check for understanding

Display the *CFU* slide and, using MWBs, ask questions like the ones below.

- Fill in the gaps for this scenario.
- What is the power of the microwave?
- How long was it switched on for?

Use this to construct a range of examples for students to attempt on MWBs before their independent practice. Include unit conversions for J, W and s. Do not allow students to progress to the independent practice until they get questions like these correct. If necessary, use more examples with different values.

Independent practice review

Student Practice Book questions 114–21 (page 85).

Q	Notes
115b	Ensure that when students see 60 s they do not convert it out of instinct back into minutes.
120	Whichever does it faster. This question will demonstrate those with a deep conceptual understanding.

Guided explanation

How can we predict what happens to the energy transferred?

Write the following power equation on the board, accompanied by an explanation, similar to the one that follows.

energy transferred = power x time

$E = P \times t$

1 Ask students what will happen if the power of the device is increased but the time is kept constant. Draw an upward-pointing arrow by the *P*.

2 Ask if we can predict what happens to the energy transferred.
3 Explain that since the power is the number of joules transferred in one second, then if the power is increased, the energy transferred will also increase. Draw an upward-pointing arrow by the *E*.
4 Repeat the process, but this time ask what will happen if the power of the device is *decreased* but the time is kept constant. Draw a *downward*-pointing arrow by the *P*.
5 Ask if we can predict what happens to the energy transferred.
6 Explain that since the power is the number of joules transferred in one second, then if the power is decreased, the energy transferred will also decrease. Draw a *downward*-pointing arrow by the *E*.

Check for understanding

Display the *CFU* slide and, using MWBs, ask questions like the ones below.

- What happens to the energy transferred if the power is increased?
- What happens to the energy transferred if the power is decreased?
- What happens to the energy transferred if the time is increased?
- What happens to the energy transferred if the time is decreased?
- What happens to the energy transferred if the power is doubled but the time is halved?

Independent practice review

Student Practice Book questions 122–29 (page 86).

Q	Notes
129d	The answer to this is exploratory because this has not been explicitly taught, but it does provide a bridge to the next topic.

P1.4 Energy resources

Students' learning objective: *I am learning about energy resources so I can understand where the energy we use in daily life comes from.*

We obtain the energy needed to generate electricity and run our vehicles from a number of sources. These sources are called *energy resources*, to distinguish them from the energy stores we have discussed in the previous topics.

Knowledge check

After working through the *General science quiz*, display the *Knowledge check* slide and, using MWBs, deliver the questions one by one, reteaching concepts as necessary.

	Question	Answer	Notes
1	What kind of energy store is a fuel?	Chemical energy store	Encourage use of the word 'store'.
2	How is energy released from a fuel?	The fuel is burned	
3	What kind of energy store is a moving object?	Kinetic energy store	Encourage use of the word 'store'.
4	What transfer is involved when an object falls?	Transfer by a force	Accept 'mechanical working'.
5	What transfer is involved in an electric circuit?	Transfer by electric current	

Guided explanation

Where do we get our energy from?

Introduce the topic by explaining that our current lifestyles mean that we cannot rely on just the energy we get from food. Display the first slide and click through the images which show the different things we require energy for. Say that, because energy cannot be created or destroyed, we must obtain 'extra' energy from resources such as fuels and mains electricity supplies.

Draw the diagram below sequentially, with each step accompanied by an explanation, similar to the one that follows.

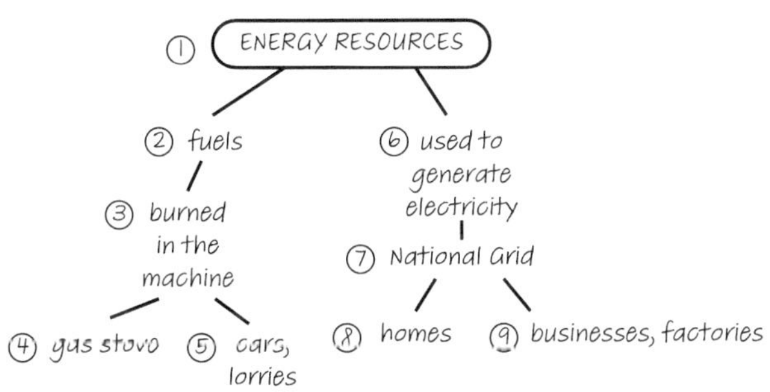

1 These additional sources are called *energy resources*.
2 Energy resources such as fuels can be used directly in machines.
3 Fuels are burned in machines.
4 For example, natural gas is burned in a gas stove.
5 Petrol and diesel fuels are burned in the engines of vehicles.
6 However, other energy resources are used to generate electricity.
7 The energy from them is transferred to our homes by electric current through the wires and cables of the National Grid.
8 The National Grid is the system of cables that transfers energy to homes.
9 It also transfers energy to businesses and factories.

Check for understanding

Display the *CFU* slide and, using MWBs, ask questions like the ones below.

- Name at least one machine that transfers energy by burning fuel directly.
- Name at least one machine that transfers energy by being plugged into mains electricity.
- A house on a remote island is not connected to the National Grid. What does this mean and how would this affect your lifestyle if you lived there?

Independent practice review

Student Practice Book questions 130–38 (page 87).

Q	Notes
130	Ensure students give multiple answers.
134	Students might say that they are not sure what a steam engine is, but the question can be answered without that knowledge since students just need to think of portable fuel options.

Guided explanation

What are fossil fuels?

Start by describing the scenario of using a wood fire to cook on, for example to toast a piece of bread. Draw the diagram below sequentially (from right to left), with each step accompanied by an explanation, similar to the one that follows.

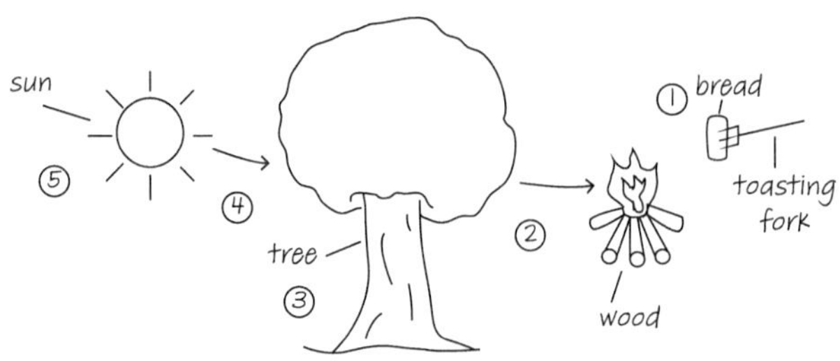

1 The wood is a chemical energy store.
2 We know that energy cannot be created or destroyed, so the energy must have come from a source.
3 Wood comes from trees.
4 But where did the tree get its energy from?
5 The tree originally got its energy from the Sun.

To highlight the concept, you can add indicative figures to the diagram, e.g. you could add '1000 kJ' to the arrow at (4), then add the same number into the tree, and then a smaller number into the pile of wood that is harvested from the tree. The main point is to emphasise that the energy starts with the Sun.

Where does the energy in fossil fuels come from?

The next part of the explanation looks at the origin of coal. If the diagram does not fit on your board in one row, curve it down and round to show the overall process. The key point throughout the explanation is to emphasise the origins of the items – both the energy and the fuel.

Tell students you are going to link wood with how a so-called 'fossil fuel' like coal gets its energy. Draw the diagram at the end of this box sequentially (from right to left), accompanied by the explanation below.

1 The coal is a chemical energy store. (If you have a piece of coal, it would be good to show students.)
2 We know that energy cannot be created or destroyed, so its energy must have come from somewhere.
3 Coal comes from a coal mine. The coal is often found deep underground. How did it get there and how did it get so much energy?
4 The process of making coal takes millions of years.
5 It is made from the remains of dead trees that became covered with mud and sand. The mud and sand became compacted (squashed) and turned into rock.
6/7 The trees were originally alive. They grew in the same way we saw in the previous diagram …
8 … and obtained the energy they needed to grow …
9 … from the Sun.

Explain that coal is called a fossil fuel because it is a fuel made from a fossil – an ancient living organism that existed millions of years ago.

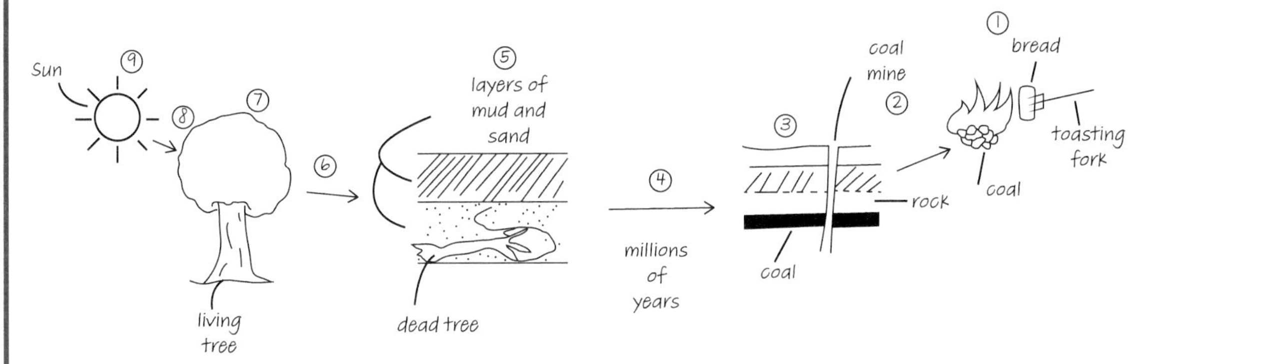

Check for understanding

Display the *CFU* slide and click to reveal the different parts of the diagram. Using MWBs, ask questions like the ones below, adding labels to the slide as you go.

- What type of energy store is coal?
- Where do we find coal?
- How long does it take for coal to form?
- Where does coal originally come from?
- What needs to happen before trees will turn into coal?
- Where did the trees originally get their energy from?
- Explain how the energy in coal dug up today originally comes from the Sun millions of years ago.
- Explain why coal is called a fossil fuel.
- Explain why wood is not called a fossil fuel.
- Explain why dinosaur bones are not called fossil fuels.

Independent practice review

Student Practice Book questions 139–49 (pages 87–88).

Q	Notes
149	If students have not completed lessons on photosynthesis, leave this question.

Guided explanation

How do we generate electricity?

Tell students that there are many ways of generating electricity and we are going to look at three that are commonly used. We will look first at how they work; then we will look at their advantages and disadvantages.

1 Display the first slide, which shows some wind turbines. Explain that a wind turbine is a very large fan on a pole. Ask students how it generates electricity.

2 Draw the following diagram to explain how a turbine and generator work. When something makes the turbine spin, the turbine turns a generator. The generator takes this turning motion and generates an electric current. (Students do not need to know any more than *the turbine turns a generator and the generator generates an electric current.*).

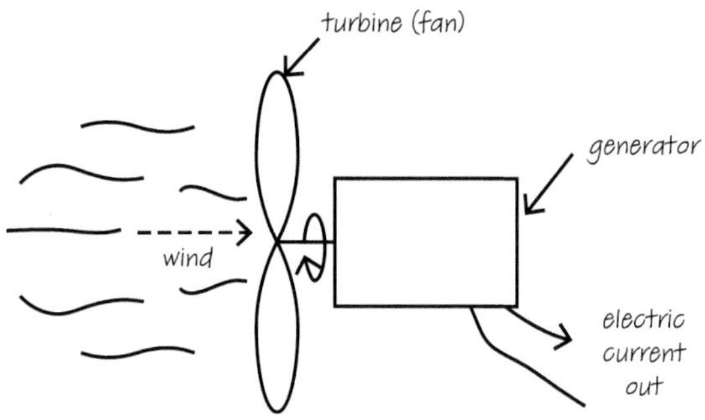

3 If you have a small hand-cranked generator available, demonstrate its use to the class.

4 Explain that a generator is a machine that uses movement to generate electric current. (Students do not need to know more detail than that at this point.)

5 Display the third slide, which shows a hydroelectric power station.

6 Draw the following diagram to show how a hydroelectric power station works.

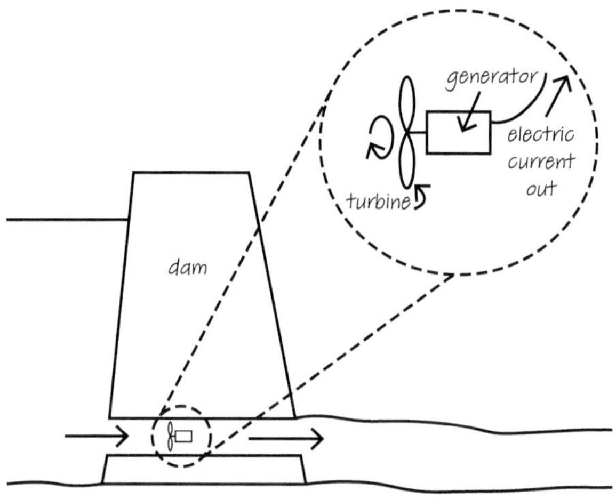

7 Explain that the water trapped behind the dam goes through a small channel. As it goes through, it makes a turbine turn. The turbine is connected to a generator, which produces electric current.

8 Display the next slide, which shows a large fossil fuel power station.

9 Draw the following diagram to show how a fossil fuel power station works.

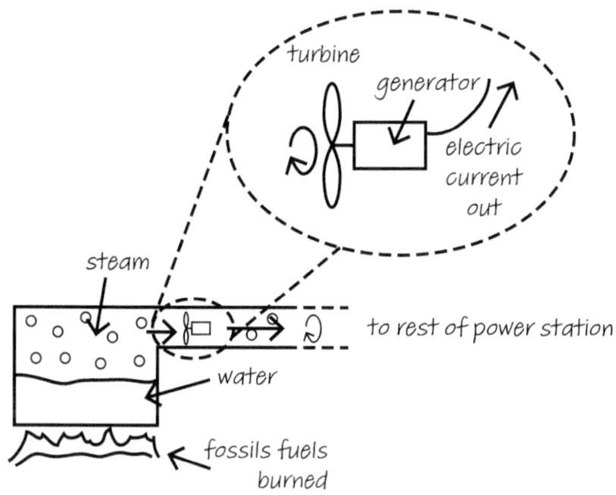

10 Explain that fossil fuels are burned and used to boil water. The steam goes through a pipe. As it goes through, the steam turns a turbine. The turbine is connected to a generator, which generates electricity.

Check for understanding

Display the *CFU* slide and, using MWBs, ask questions like the ones below.

- In a wind turbine, does the 'fan' move the wind, or does the wind move the 'fan'? *During KS3, teachers are advised to focus on the physical processes involved in the methods of generating electricity rather than descriptions of changes in energy stores. Energy stores will be useful at KS4 when calculations involving efficiency and so on will be considered.*
- In a hydroelectric power station, what substance is used to turn the turbine?
- In a fossil fuel power station, what substance is used to turn the turbine?
- In all the methods for generating electricity considered, what is the name of the machine that produces electric current?

Independent practice review

Student Practice Book questions 150–58 (page 88).

Q	Notes
150	'Turbine turns a generator and a generator generates an electric current' is sufficient.
155	The time is not mentioned because the time to fall is not the relevant quantity. The relevant time would be how quickly energy is transferred to the turbine, but this has been omitted to avoid causing confusion. We can simplify here to 'more energy transferred means more power' (we assume the same time to transfer).

Guided explanation

What is the best way of generating electricity?

Display the table on the slide and use it to summarise how each method of electricity generation works, adding details as shown below. Use the opportunity for *Cold Call* questioning to check that students understand each method.

Energy resource	Diagram	Advantages	Disadvantages
wind power		• wind is free ① • wind replaced naturally each day • does not produce carbon dioxide ⑧	• turbines expensive to build • may not be windy enough • noisy ②
hydroelectric	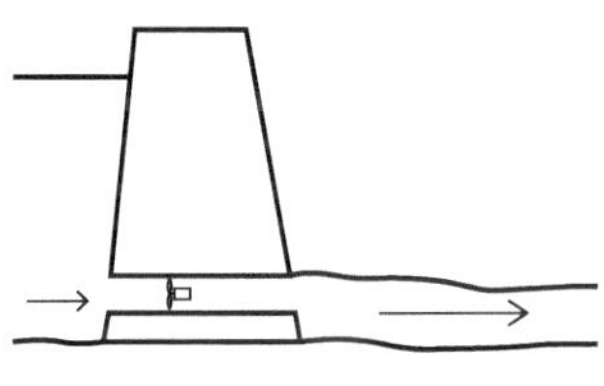	• water is free • replaced naturally by rainfall ③ • does not produce carbon dioxide ⑨	• expensive to build • only a few locations are suitable • may be affected by drought ④
fossil fuel power station	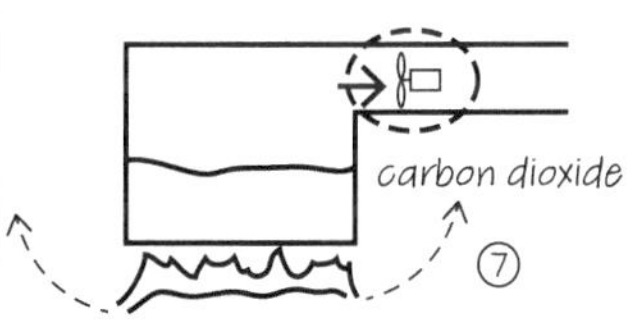	• not affected by weather ⑤ • can work on a very large scale	• fuel can be expensive • fossil fuels not replaced naturally • produces large amounts of carbon dioxide ⑥

1 Consider the advantages of using wind turbines to generate electricity.
2 Then consider the disadvantages.
3 Consider the advantages of using hydroelectricity to generate electricity.
4 Then consider the disadvantages.
5 Consider the advantages of using fossil fuels to generate electricity.
6 Then consider the disadvantages.
7 Fmphasise that burning fossil fuels creates large amounts of carbon dioxide, which contributes to global warming.
8 Add to the advantages of wind power that it does not produce large amounts of carbon dioxide.
9 Add to the advantages of hydroelectric power that it does not produce large amounts of carbon dioxide.

Check for understanding

Display the *CFU* slide and, using MWBs, ask questions like the ones below.

- Which methods of generating electricity have zero cost for fuel?
- Which method of generating electricity is not affected by weather?
- Which method of generating electricity creates a gas that has an impact on the environment?
- What is the name of the gas?
- Which energy resources are replaced naturally over a short time scale? *Fossil fuels are replaced naturally, but on a time scale of millions of years.*

Independent practice review

Student Practice Book questions 159–72 (pages 88–89).

Q	Notes
164	This is a precursor to the next section on renewable and non-renewable energy. Wind and water are resources that are easily replaced, whereas fossil fuels are not.

Guided explanation

What does renewable mean?

Tell students that we can sort different energy resources into groups. We could choose to sort them by 'which energy resources use aspects of the weather' or 'which energy resources involve water'. (These are good opportunities to check for understanding.)

Say that one common way to sort energy resources is to look at whether we will still be able to use the method in the future. Draw the diagram below sequentially, accompanying each step with an explanation, similar to the one that follows.

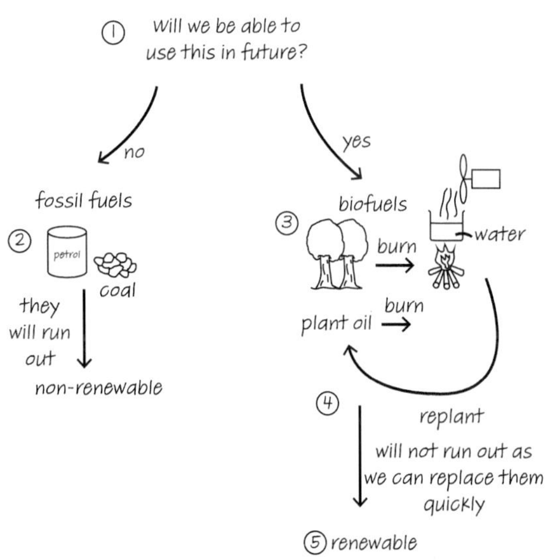

1. We know that we will run out of fossil fuels eventually. This is because we are using them up incredibly quickly and they cannot be replaced.
2. We say that fossil fuels are *non-renewable* – we will run out of them eventually.
3. We know that some energy resources will never run out. Biofuels are fuels made from biological materials. We grow them, then burn them or their oils. But we can replant new crops to replace them relatively quickly.
4. We call this a *renewable resource.* This gives us our best definition of renewable – something that can be replaced as quickly as we use it.
5. Renewable does not mean that we can use it again. We cannot use the energy from the Sun again once it has hit a solar panel, but there will be more light from the Sun for us to use.

Use the flow diagram on the second slide to work through a few examples. When using wind or solar power, we do not 'use up' the Sun or the wind. We call these renewable too, since we will not run out of them, and we are not burning any fuels on Earth with these resources.

Check for understanding

Display the *CFU* slide and, using MWBs, ask questions like the ones below.

- Give one example of a renewable energy resource. *Accept any renewable energy resource, e.g. solar, tidal, wave, biofuel.*
- Why are some energy resources called *renewable*? *Discourage 'They will not run out' because this is dependent on them being naturally replaced.*
- Is wood a renewable energy resource? *Scope for discussion here: wood is renewable only if new trees are planted to replace any that are used.*
- Is solar power a renewable energy resource? *Sunlight is replaced naturally.*
- Are waves a renewable energy resources?
- Give one example of a non-renewable energy resource. *Accept a named fossil fuel, e.g. coal, oil, gas. Accept nuclear fuel.*
- Why are some energy resources called *non-renewable*? *Discourage 'They will run out' because this is dependent on how quickly they are used. Non-renewable resources could in theory last hundreds or thousands of years with careful management.*

Independent practice review

Student Practice Book questions 173–78 (page 89).

Q	Notes
175	In case students mention it, nuclear power is non-renewable because we will eventually run out of nuclear fuel, and it cannot be replaced.

P2 Speed

P2.1 Speed

Students' learning objective: *I am learning about speed so I can calculate how fast different objects travel.*

Speed was mentioned briefly when we discussed the kinetic energy store in Topic P1.2. It is a concept that students will have some familiarity with from everyday life, and this topic will introduce speed in a mathematical context – how to calculate it, measure it and read it from a graph.

Knowledge check

After working through the *General science quiz*, display the *Knowledge check* slide. The questions are best delivered one by one on mini-whiteboards (MWBs). If students do not know the answer to any of them, you should reteach the relevant concept. Answers are not on the slides because it is best to go over them together as a class.

Questions 1–5 are to check whether students have the procedural knowledge of how to perform a calculation and a unit conversion. If you find that students are particularly adept at these, then you may want to skip ahead. Questions 6–8 are to gauge understanding of knowledge from KS2 and outside school.

	Question	Answer	Notes
1	How much energy is transferred by a 9 W lightbulb, left on for 5 s?	45 J	
2	A 500 W microwave is left on for 2 s. How much energy is transferred?	1000 J	
3	A 200 W TV is left on for 2 minutes. How much energy does it transfer?	24 000 J	Ensure students have converted the time from minutes to seconds.
4	How much energy is transferred by a 2 kW heater, left on for 8 s?	16 000 J	Ensure students have converted kilowatts to watts.
5	I leave on a 1.5 kW kettle for 3 minutes. How much energy is transferred?	4500 J	Ensure students have converted both the time and the power.
6	What does speed mean?	How fast something travels	
7	What does stationary mean?	Not moving	You may need to show the difference between 'stationary' and 'stationery'.
8	What units do we use to measure speed in in everyday life?	Miles per hour/ kilometres per hour	Look out for answers like 'mph' and 'kph'. You will need to address how the 'p' is used in everyday life versus '/' in science.

Guided explanation

What does speed mean in a race?

Introduce the topic with an example. Display the diagram and table on the slide, which show the results of a running race. Allow students some time to quietly process them. Then call their attention back to you and give an explanation, similar to the one below.

Racer	Distance (m)	Time (s)
Tom	100	12
Ayaan	100	9
Claire	100	11
Maleeka	100	14

1 If we think of a race, the runners all travel the same distance and we time how long it takes them. Faster racers will cover the distance in less time.

2 We know that Ayaan is the fastest because he covered the distance in the least time, so his speed is the highest. We know that Maleeka is the slowest because she covered the distance in the most time, so her speed is the lowest. We know that Claire is faster than Tom, so her speed is higher than Tom's because she covered the distance in less time than him.

Check for understanding

The Check for understanding (CFU) is best conducted using MWBs. Display the *CFU* slide and ask questions like the ones below. Tailor and adapt your questions to the class in front of you.

- Who is the fastest in the race?
- How do you know? *Do not accept just 'X was quickest'. We want students to see that fast things cover the distance in less time.*
- Who is the slowest in the race?
- How do you know? *As above – here students should refer to covering the distance in more time.*
- Who is faster: Rahees or Taylor?
- How do you know?

There is no independent practice at this stage. If you think your class needs some consolidation, ask further questions of a similar style to the CFU.

Guided explanation

What does miles per hour mean?

Display the table on the first slide. Explain that in real life, the occasions that we need to calculate speed are not simple races. It is more like we want to know who is fastest when our racers have all run different distances in different amounts of time. How can we figure out which one of these racers is fastest now? And what if we only have one object? How do we know if it is fast or slow?

Explain that we can measure how fast an object is by calculating or measuring its speed. Display the second slide and annotate it sequentially, as shown below. Accompany each step with an explanation, similar to the one that follows.

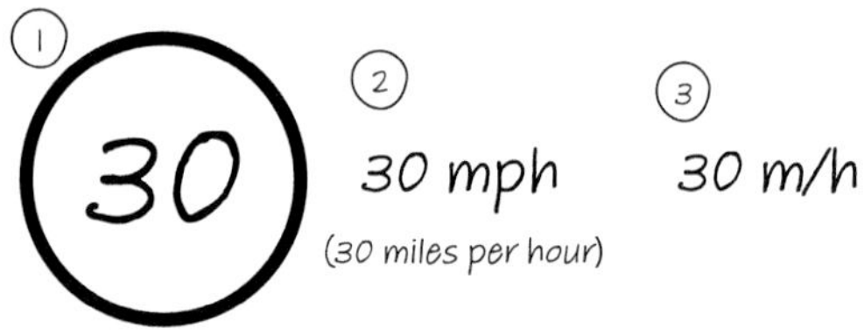

1 All speeds have a unit. Different countries have different accepted units for speed when it comes to vehicle speeds on roads. In the UK, our road speeds are given in miles per hour.
2 What the number 30 in the red circle means is the number of miles that are travelled in a specific amount of time – 1 hour. We write this as 30 mph. So 30 mph means that an object will travel 30 miles in each hour – that is its speed. 50 mph means that an object will travel 50 miles in each hour – that is a faster speed.
3 In everyday life we often write 'miles per hour' as 'mph' but in science, the 'p' for per has a mathematical symbol. Per means divide. So we can write 30 mph as 30 m/h instead. We can see how fast an object is going now without comparing with other objects – just by looking at its number for speed.

Check for understanding

Using MWBs, ask questions like the ones below.

- How far will you travel in an hour if you are walking at 4 miles per hour? *Ensure that students give units with their answers.*
- How far will you travel in an hour if you are driving at 60 miles per hour?
- What is the mathematical symbol for 'per'?
- Display the *CFU* slide and ask: How far will you travel in 2 hours if you are travelling at 50 miles per hour? *This question acts as a link into the next section.*

There is no independent practice at this stage. If you think your class needs some consolidation, ask further questions of a similar style to the CFU.

Guided explanation

What do we measure speed in?

Draw the diagram below sequentially, with each step accompanied by an explanation, similar to the one that follows.

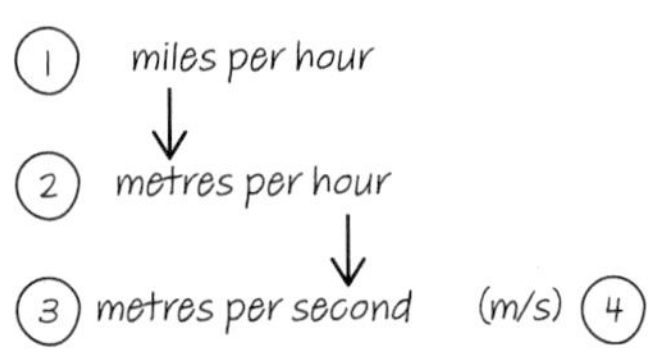

1 In science, we do not use miles per hour as our units. This is because in science we do not use miles to measure distances and we do not use hours to measure time.
2 In science, we use metres to measure distance. Therefore, instead of miles, we should use metres. So metres per hour is more scientific than metres per hour.
3 In science, we use seconds to measure time. Therefore, instead of hours, we should use seconds. So metres per second is even more scientific than miles per hour. Our standard units of speed are metres per second.
4 We write 'metres per second' as m/s. We read the slash (the divide sign) as per. Therefore, in science, we will talk about objects travelling at speeds like 4 m/s. This means that the object will travel 4 m in each second.

Check for understanding

Using MWBs, ask questions like the ones below.

- In science, what do we measure speed in? *Check that students know the m is metres and not miles.*
- What does the / mean in m/s?
- Two cars are in a race. One is travelling at 10 m/s and the other at 11 m/s. Which is faster?
- How do you know?

Display the *CFU* slide and ask questions like the ones below, clicking to reveal each question one at a time.

- I am running at 6 m/s. How far do I travel in each second? *Ensure that students include units.*
- I am running at 6 m/s. How far do I travel in 2 s? *Ensure that students include units. Here we are testing students' understanding of 'per'.*
- I am running at 6 m/s. How far do I travel in 10 s? *Ensure that students include units. Here we are testing students' understanding of 'per'.*

There is no independent practice at this stage. If you think your class needs some consolidation, ask further questions of a similar style to the CFU.

Guided explanation

What does speed mean?

Display the slide, which shows the speed of four different types of vehicle and a person. Work through them one at a time, accompanied by an explanation, similar to the one below.

1 If we measure how far a bus travels in 1 s, it might travel 5 m. We say the bus travels at 5 m/s as this means it goes 5 m in every second.

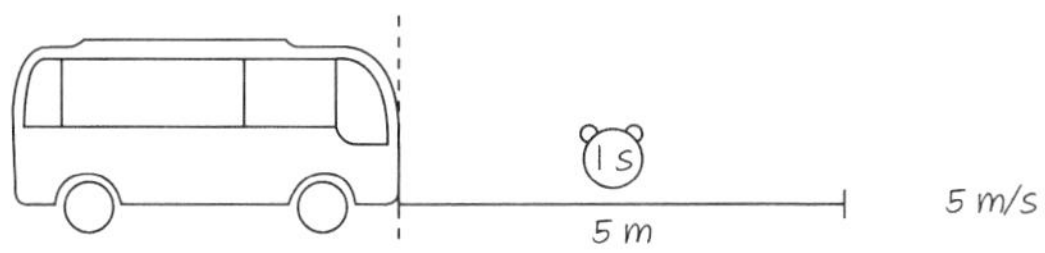

2 If we measure how far a bicycle travels in 1 s, it might travel 3 m. We say the bicycle travels at 3 m/s as this means it goes 3 m in every second. The bicycle is slower than the bus because it travels less far each second.

3 Repeat this for the person, car and train.

Explain that speed is simply how far an object travels in 1 second of time. The further the object travels in 1 second, the faster it is moving.

Check for understanding

Display the *CFU* slide and, using MWBs, ask questions like the ones below.

- Which object is fastest?
- How do you know?
- How fast is the train travelling? *Ensure that students include units.*
- Which object is travelling at 5 m/s?
- How do you know?
- Which object is slower than the bicycle?
- How do you know?

Independent practice review

Students practise independently by answering questions 1–14 from their Practice Books (pages 90–92). Full answers can be found on the *IP answers* slides or in the answer documents on Boost. Notes are only included for questions, where useful.

Q	Notes
5b	When asking 'how do you know' about which object is faster, ensure that students are thinking about the definition of speed, and not just which number is bigger. 5 m/s is fastest because it travels 5 m each second, whereas the others travel less far each second.
7c, d	Questions of this type already make reference to the object, so it is fine for students to just say 'chemical energy store' rather than 'chemical energy store of the dog'.
11d	This is in addition to the original 200 J, so if students just answer 500 J, they have not added the new amount to the original amount.
12	Ensure that students realise that the / means divide.
14g	Answers here have the opportunity for quite a lot of detail, explicitly linking the speed to the kinetic energy store, the kinetic energy store to the gravitational potential energy store, and the gravitational potential energy store to the height.

Guided explanation

How do we calculate speed?

Draw the diagram below sequentially, with each step accompanied by an explanation, similar to the one that follows.

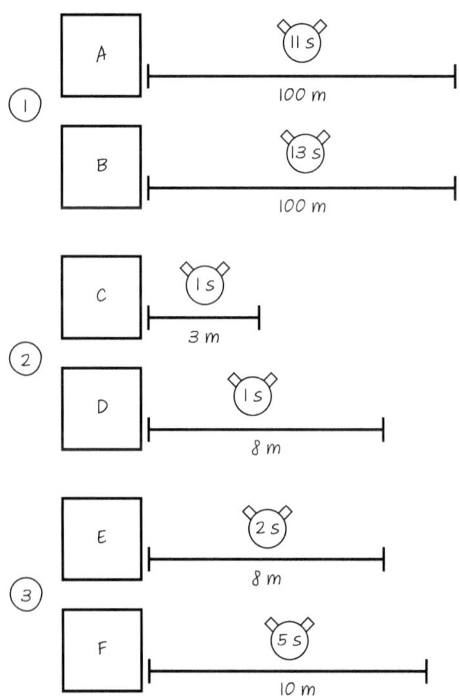

1 If we know that two objects have travelled the same distance, it is easy to work out which is faster – we look for the one that covered the distance in the least time. Here A is faster because they covered 100 m in less time than B.

2 If we know that two objects have travelled for the same amount of time, it is easy to work out which is faster – we look for the one that travelled the furthest. Here D is faster because they travelled a further distance in 1 s than C.

3 But what if we have two objects that have travelled different distances and taken different amounts of time – how do we work out which is faster out of E and F?

Draw the diagram below sequentially, with each step accompanied by an explanation, similar to the one that follows.

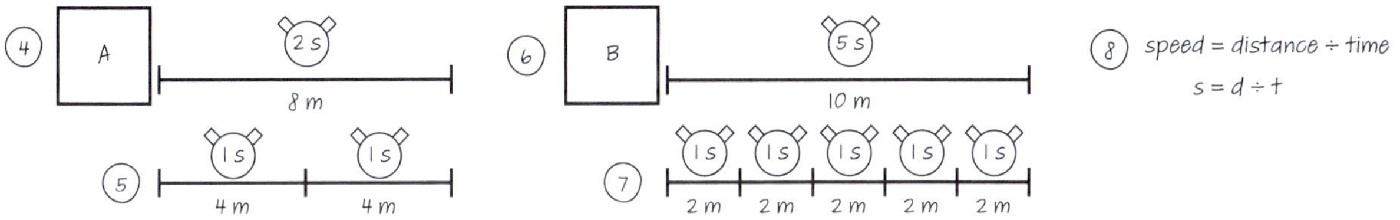

4 Object A has travelled 8 m in 2 s. We want to work out the speed, which is the number of metres travelled in each second.

5 We can see that it travelled 8 m in 2 s, so it must have travelled 4 m in each second, i.e. 4 m in the first second and 4 m in the next second, to achieve 8 m in 2 s.

6 Object B travelled 10 m in 5 s.

7 We want to know how far it travelled in 1 s. It must have travelled 2 m in each second, i.e. 2 m in the first second, 2 m in the next second, and so on, to achieve 10 m in 5 s.

8 What we are doing is dividing the distance travelled by the number of seconds taken. By doing this we see the number of metres travelled in just 1 second – the number of metres per second – which is the speed. We write this equation in words and symbols like this. This is the equation for speed.

How do we use the speed equation?

To explain how to use the speed equation, go through the two worked examples on the slides, showing students how to use the EVERY method, one of which involves a unit conversion. Do the first worked example yourself, explaining what you are doing at each step.

A lorry takes 12 s to travel 48 m. What is its speed?	
Equation	speed = distance ÷ time $s = d \div t$
Values	s = ? d = 48 m t = 12 s
Enter values	$s = 48 \div 12$
Result	$s = 4$
Y(units)	s = 4 m/s

Remind students that using EVERY ensures that they show clear working out. For the second example, start verbally, asking students to fill in the next steps.

Conversions involving the kilo prefix were covered in Topic P1.3.

Check for understanding

Display each of the *CFU* slides in turn, with example scenarios for students to calculate speed.

- For example 1, ensure that students include all the steps of the working out, including the correct units.
- Example 2 will require students to use a calculator.
- For example 3, ensure that students have clearly converted the 3 km into m.

Do not allow students to progress to independent practice until they are getting questions like these correct.

Independent practice review

Student Practice Book questions 15–33 (pages 92–93).

Q	Notes
33b, c	Ensure that students are writing X energy store *of* Y, and linking the energy store to an object when it is not clear from the question what the object is.

Guided explanation

How do we deal with multiple equations?

Students have now encountered two different equations in physics – one linking speed, distance and time ($s = d \div t$) and one linking energy transferred, power and time ($E = P \times t$). In this section, you will teach students how to select the correct equation for a given scenario, and they will compare the similarities and differences between the equations.

Display the example on the first slide, which includes both equations mentioned above.

1. Explain that we cannot use any equation with any question. So how do we know which one to use?
2. Point out that the units of the values in a question tell us what physical quantities are represented, so we can use the units to help us understand what the values represent and can therefore find equations with these physical quantities in them.
3. Talk through the example, showing how the units of W means power and s means time. If we have power and time, then we can use $E = P \times t$ to calculate the energy transferred.

A remote control car's motor has a power of 800 W. It runs for 30 s. What can we calculate about the car?	
Equation	energy = power × time $E = P \times t$
Values	$E = ?$ $P = 800$ W $t = 30$ s
Enter values	$E = 800 \times 30$
Result	$E = 24000$
Y(units)	$E = 24000$ J

4. We *cannot* use $s = d \div t$ to calculate the speed though, as we do not have enough information – we do not know how far it travelled.
5. Repeat the process for the example on the second slide, this time showing that we can use the speed equation but not the energy transferred equation.

Check for understanding

Display each of the *CFU* slides in turn and, using MWBs, ask questions like the ones below.

- What can you calculate from the data? *But don't get them to calculate it yet.*
- How do you know?
- Why can you not calculate the energy transferred/speed?
- Calculate the speed/energy transferred.

Independent practice review

Selecting equations worksheet on Boost and Student Practice Book questions 34–46 (pages 94–95).

Q	Notes
38e	Discuss how the air warms up and wasted energy is transferred to thermal energy stores.
41a	3 000 000 J. Ensure students show their conversions of both power and time.
41c	Questions of this type already make reference to the object, so it is fine for students to just say 'gravitational energy store' rather than 'gravitational energy store of the block'.
41e	As before, we are assuming 100% efficiency.

Guided explanation

How do we measure speed?

Draw the diagram below sequentially, with each step accompanied by an explanation, similar to the one that follows.

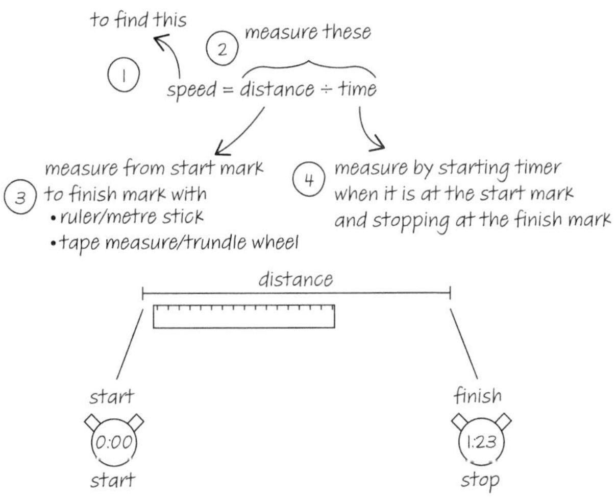

1. Let us imagine we want to measure the speed of a car from the outside. From the inside of the car, the speedometer can tell us the speed, but we cannot see that from the outside. There are some tools that we could use to measure speed directly – like a police radar gun – but these are expensive and uncommon. There is a simple way to measure the speed of an object from the outside using the equation for speed.
2. To find the thing on the left-hand side of this equation we need to measure all of the things on the right-hand side of the equation.
3. We measure the distance the object has travelled between two points – a start and finish mark – with a ruler or metre stick (for small distances), or even a tape measure or trundle wheel (for large distances).
4. We measure the time it takes the object to travel this distance by starting a timer when it hits a start mark, and stopping the time when it hits a finish mark.

Check for understanding

Display the *CFU* slide and, using MWBs, ask questions like the ones below.

- What two things do we need to measure to find the speed of an object?
- Where do we measure the distances from and to?
- What can we use to measure small distances?
- What can we use to measure large distances?
- What can we use to measure the time taken?

Independent practice review

Student Practice Book questions 47–52 (page 95).

Q	Notes
50	Ensure that students link the quantity to be measured with the equipment, e.g. 'measure the distance travelled using a ruler'.

Core practical

Measuring speed: Part 1

A method is provided on the *Practical worksheet* and *Teacher and technician notes* on Boost – including equipment list and safety notes – if you wish to allow your students to proceed independently. However, for most students it will ordinarily be better to take a *Slow Practical* approach following the guidance below.

In this practical, we will introduce a crucial aspect of practical science – *lines of inquiry*. To do this, we are going to set up the practical and take a measurement of speed and then discuss what *lines of inquiry* are, before returning to the practical later in the topic to apply this.

Students should work in pairs. You and each pair will need a toy car (or ball), wooden ramp, metre rule, stopwatch, masking tape (or similar), clamp stand and G-clamp (or masses).

1. Set up a demo station at the front. Assign the equipment by getting students to collect apparatus as needed and return to their workstations.
2. Demonstrate how to set up the ramp so that one end is raised up on the arm of a clamp stand – at this point do not worry about getting the height of the ramp to an exact value. Ask students to set up their equipment. Ensure the ramp is safely secured.
3. Demonstrate how to set up the start mark and end mark on the ramp with a piece of masking tape. Tell students that to find the distance travelled, we must measure the distance between the tapes. Discuss whether you will measure from the start or end of the pieces of tape.
4. Ask students mark their start and finish lines, and measure the distance the car will travel each time.
5. Demonstrate how to release the car/ball and start/stop the stopwatch at the appropriate moments. Ask students to take their first measurement of the time taken.
6. From the distance travelled and the time taken, ask students to calculate the speed of their cars using the EVERY method, as they have done previously. (A copy of the EVERY table is provided on the second slide.)
7. For independent practice on this part of the practical, ask students to answer the questions in the first section of the *Practical worksheet*.

Here we are going to pause the practical to discuss lines of inquiry, variables, predictions and the drawing of tables.

Guided explanation

What is a line of inquiry?

Draw the diagram below, in the context of the previous practical. Construct it sequentially, with each step accompanied by an explanation, similar to the one that follows.

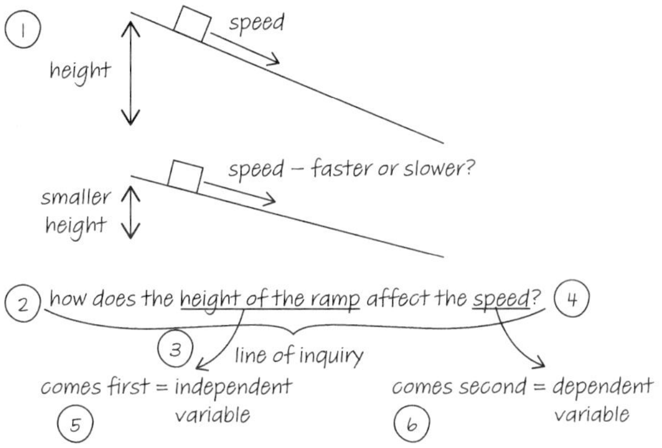

1 As scientists, we want to know what happens if we change the height of the ramp, and make it smaller. Would it make the car faster or slower? This is probably obvious, but how do we actually work it out?
2 What we want to know, then, is how does the height of the ramp affect the speed?
3 Scientists ask questions of this type all the time – we call these types of question *lines of inquiry*.
4 Each line of inquiry (LOI) has two halves – in our example we have the height of the ramp and the speed.
5 The thing that comes first in the LOI – in our case, the height of the ramp – is called the *independent variable*. We will not go into depth about the reason why it is called this right now.
6 The thing that comes second in the LOI – in our case, the speed – is called the *dependent variable*. Again, we will not go into depth as to the reason for this right now.

Now look at some other lines of inquiry. Write the following on the board, accompanied by the following explanations.

(1) How does the power of a heater affect the temperature of a room?
? ?

(2) The power of a heater affects the temperature of a room.

1 If we have a heater in the room, we might ask: 'How does the power of the heater affect the temperature of the room?' This LOI has two parts. The first part – the part doing the affecting or the changing, i.e. the power of the heater – is the independent variable. The second part – the thing being affected or changed as a result, i.e. the temperature of the room – is the dependent variable.
2 This second example is *not* an LOI. It is not phrased as a question. We are not asking how one thing might be linked to another, we are just saying that they are linked. This is not a scientific exploration, so we do not call it an inquiry.

Draw the experimental set up below, accompanied by the explanation that follows.

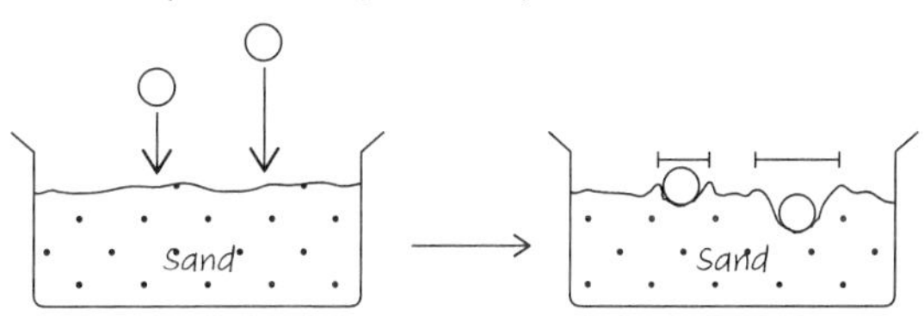

Explain the set-up of dropping two balls into a tub of sand from different heights, and looking at the widths of the craters that are produced. Use this to begin to question students about the LOI.

- What is the cause?
- What is the effect?
- How do we write the LOI?
- Why is it not 'How does the size of the crater affect the height?'
- What is the independent variable?
- What is the dependent variable?

Ensure that students see that the LOI is written as '*How does the* independent variable *affect the* dependent variable?'

Check for understanding

Display the first *CFU* slide and, using MWBs, ask questions like the ones below.

Before asking the questions, describe the context of the experiment: *I have a beaker of water, and I add an ice cube to it. I time how long it takes for the ice to melt. I am interested to know how the temperature of the water affects how long it takes for the cube to melt.* Then click to reveal the line of inquiry (LOI).

- Which is the independent variable?
- How do you know? *Accept that it is the cause of the change.*
- Which is the dependent variable?
- How do you know? *Accept that it is the thing being changed.*

Display the second slide and ask questions like the ones below.

Before asking the questions, describe the context of the experiment: *I know that doing exercise will increase my heart rate. I am going to run for different lengths of time and then measure my heart rate after each run. I am interested to know how the amount of time that I run for affects my heart rate.* Then show the LOI.

- Is the heart rate the dependent or independent variable?
- How do you know?
- Is the time that you have run for the dependent or independent variable?
- How do you know?

Display the third slide and ask questions like the ones below.

Before asking the questions, describe the context of the experiment: *I have a piece of string that I know the length of. I am going to tie knots in the string and measure the new length.*

- What is the LOI for this experiment?
- Which is the independent variable?
- How do you know?

Independent practice review

Student Practice Book questions 53–59 (pages 95–96).

Note: These concepts can be difficult for students to consolidate. If you need to, use examples from the independent practice as part of your class discussion to build their understanding.

Q	Notes
57a	Lines of inquiry must be in the form of a question – ensure that students' answers focus on this.

Guided explanation

What is a controlled variable?

Draw the diagram below sequentially, with each step accompanied by an explanation, similar to the one that follows.

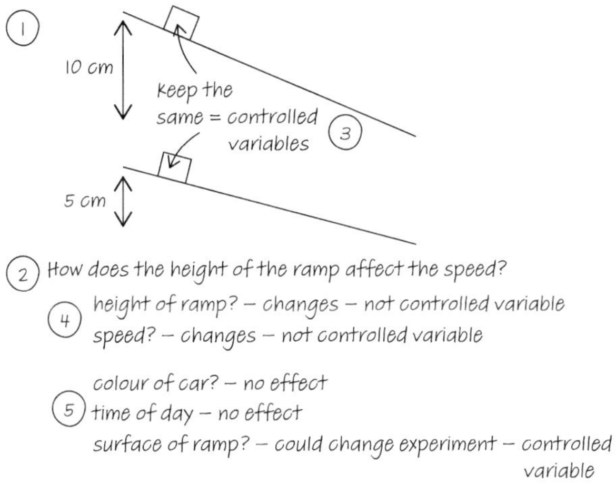

1 Returning to our speed experiment, there is another thing we have to consider in an experiment. We are looking at how the height of the ramp affects the speed of a car. We could reduce the height of the ramp and measure the speed, but we have had to use a larger car this time, as we lost the original.

2 Our line of inquiry is the same as before. But is what we have done okay? Is it okay to change to a larger car? No. We have changed something else in the experiment apart from the height. Therefore, if the speed of the car is different with a lower ramp and a bigger car, we will not know what caused the change – the ramp or the car.

3 These additional things that we must keep the same in an experiment are called *controlled variables*. We cannot change them during the experiment as they might affect the result.

4 The height of the ramp is not a controlled variable as this changes in the experiment – it is the independent variable, and we change it. The speed of the car is not a controlled variable either because it changes in the experiment – it is made to change by the ramp changing height.

5 Some things would have no effect on the car at all, like the colour of the car or the time of day that we do the experiment, so these are not controlled variables. But something like the surface of the ramp could alter our results if we changed it during the experiment, so this is a controlled variable.

Check for understanding

Display the first *CFU* slide and, using MWBs, ask questions like the ones below.

- Is the volume of water a controlled variable?
- How do you know? *Accept specific answers such as: 'Because if there was more/less water it might melt the ice faster/slower.'*
- Is the number of ice cubes a controlled variable?
- How do you know? *Accept specific answers.*
- Is the colour of a desk a controlled variable?
- How do you know?
- Is the temperature of the water a controlled variable?
- How do you know?
- Is the time taken a controlled variable?
- How do you know?

Display the second slide and ask questions like the ones below.

- What is the independent variable?
- What is the dependent variable?
- Give an example of a controlled variable.
- Is the temperature of the room a controlled variable?

- Why not?
- Is the height of the ball a controlled variable?
- How do you know?
- Is the width of the crater a controlled variable?
- How do you know?

Display the third slide and ask questions like the ones below.

- Why is the initial length of string a controlled variable?
- Give an example of another controlled variable.
- Is the ruler a controlled variable?
- Why not?
- Is the number of knots a controlled variable?
- How do you know?

Independent practice review

Student Practice Book questions 60–68 (pages 96–97).

Note: These concepts can be difficult for students to consolidate. If you need to, use examples from the independent practice as part of your class discussion to build their understanding.

Q	Notes
60	Do not accept 'things we control' (as we are also in control of the independent variable).
62	Do not accept 'so that it is fair', but get students to develop this idea of what fair means in science (we can link cause and effect).

Guided explanation

How do we write a prediction?

Use the context of the speed practical to introduce making predictions. Write the following notes on the board sequentially, with each step accompanied by an explanation, similar to the one that follows.

(1) LOI

How does the height of the ramp affect the speed of the car? (IV: height of the ramp; DV: speed of the car)

(2) Prediction

(3) (4)

As the height of the ramp increases, the speed of the car increases.

(5) *As the height of the ramp changes, the speed of the car changes.*

As the height of the ramp increases, the speed of the car changes.

(6) *As the height of the ramp increases, the speed of the car decreases.*

1. We can use our line of inquiry (LOI) to write predictions.
2. A prediction is when we are specific about what we think we will see in the experiment.
3. To write a prediction, we reword the LOI. We start by saying 'As the independent variable ...' and then we are specific about the way in which it changes – either increases or decreases. Here we will say 'As the height of the ramp increases ...'
4. We then predict how this change in the independent variable will alter the dependent variable – this is an educated guess. But as with the independent variable, we have to be specific about the way in which it changes, so we cannot just say 'the speed of the car changes'. So we will say 'As the height of the ramp increases, the speed of the car increases'. This is a prediction.
5. We can write things that might look like predictions but are not. If we are not specific about the changes, and we just say 'As the height of the ramp changes ...', then we are not actually predicting anything at all.
6. We could equally have said 'As the height of the ramp increases, the speed of the car decreases'. This is an equally valid scientific prediction (even if we know it is probably wrong), because we have been specific.

Check for understanding

Display each of the *CFU* slides in turn and, using MWBs, ask questions like the ones below.

- (Display the first slide.) Write a prediction for this line of inquiry (LOI).
- (Display the second slide.) Write a prediction for this LOI.
- (Display the third slide.) Which of the predictions 1–4 are valid predictions? Why is 1 not a valid prediction?

Independent practice review

Student Practice Book questions 69–72 (page 97).

Q	Notes
70	Focus on an LOI being formed as a question with no immediate answer, whereas a prediction is making a definitive statement that the link exists, and will work in this way.

Guided explanation

How do we draw a results table?

Continue with the context of the speed practical and explain how to draw a results table. Write the following notes on the board sequentially, with each step accompanied by an explanation, similar to the one that follows.

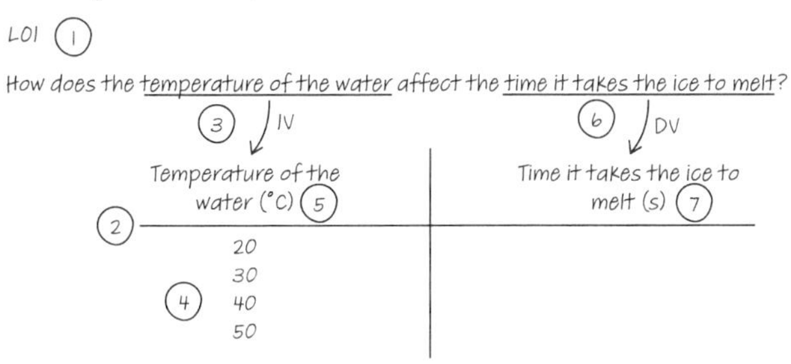

1 We can use our line of inquiry (LOI) to draw sensible tables for our experiments.
2 First, we take a pencil and ruler and draw a large 't' shape. The headings will go above the horizontal line.
3 On the left side of the vertical line, in our *first* column, we write the heading for the independent variable (the *first* part of the LOI) copied exactly from our LOI.
4 As we get to choose this, we can immediately write the numbers in that we are going to use for this. This is another big clue about which is the independent variable. If we can write the numbers into the column before we start, it must be the independent variable.
5 We do not want to write the units after every single number, so scientists put the units in the heading in brackets. By doing this, we mean '*all of the numbers in the column below have units of …*'
6 On the right side of the vertical line, in our *second* column, we write the dependent variable (the *second* part of the LOI), copied exactly from our LOI.
7 We remember to always add our units to the heading. A good clue that this is our dependent variable is that we cannot write any of these numbers into this column beforehand.

Check for understanding

Display the first *CFU* slide and, using MWBs, ask questions like the ones below.

- What are we going to write in the top left?
- Why do we write this here?
- How do we write the units?
- What units should we use for the independent variable?
- What are we going to write in the top right?

- Why do we write this here?
- Which column of numbers can we write out beforehand?

Repeat this for the second slide. Then display the third slide and ask students what is wrong with the table.

Independent practice review

Student Practice Book questions 73–78 (pages 97–98).

Q	Notes
73, 74	Ensure that students read the hint at the end of the question about the units for variables they have not seen before.

Core practical

Measuring speed: Part 2

A method is provided on the *Practical worksheet* and *Teacher and technician notes* on Boost – including equipment list and safety notes – if you wish to allow your students to proceed independently. However, for most students it will ordinarily be better to take a *Slow Practical* approach following the guidance below.

We now return to our practical on speed. We are going back to our car and the ramp, but to keep things simple we are just going to use the following line of inquiry (LOI): *How does the height of the ramp affect the time it takes for the car to move down it?* (Do not give this LOI directly to students, because they need to write this as their answer to question 1.)

Explain that we will repeat the earlier practical but now just measure the time it takes the cars to go between the marks on the ramp, and that we will alter the height of the ramp each time.

1. Display the first slide and briefly go over any practical elements again as you see fit.
2. Ask students to complete questions 1–5 on the *Practical worksheet* in which they need to write a LOI, consider the variables and write a prediction for the practical.
3. Demonstrate how to adjust the height of their ramp.
4. Ask students to construct a table on the *Practical worksheet* for their results of time taken for each height of the ramp. Remind them to include a column for speed as they will calculate this later. (A copy of the EVERY table for this is provided on the second slide.)
5. Allow students to collect their data and complete the table for a range of heights.
6. Students then answer the final question on the *Practical worksheet*, which asks them to compare their results to their prediction.

P2.2 Distance-time graphs

Students' learning objective: *I am learning about distance-time graphs so that I can calculate the speed of an object from a graph.*

We are going to introduce students to simple interpretations of graphs by applying their knowledge of speed to distance-time graphs. Our choice of examples will be important in supporting students to calculate the gradient of the line (whilst we do not make explicit references to this being the 'gradient').

Knowledge check

After working through the *General science quiz*, display the *Knowledge check* slide and, using MWBs, deliver the questions one by one, reteaching concepts as necessary.

	Question	Answer	Notes
1	Which is faster: 4 m/s or 6 m/s?	6 m/s	
2	Explain your answer to Q1.	Because it travels more distance in each second	Ask students to elaborate on 'the number is bigger'.
3	In words, what is the equation for speed?	speed = distance ÷ time	
4	A car travels 222 m in 5 s. What is its speed?	44 m/s	
5	How can we measure the speed of an object?	Measure the distance travelled and measure the time taken to travel that distance, and then use the equation	
6	Which is the horizontal axis of a graph – the *x*-axis or the *y*-axis?	*x*-axis	As this is 50:50, probe deeper and ask further questions here if necessary.
7	Give the values from this scale for letters A–D. (See diagram on slide.)	A: 20, B: 42, C: 58, D: 84	

Guided explanation

How do we show moving objects on a graph?

Introduce the topic by explaining that we draw graphs in science to compare data or to see trends. When we plot moving objects on a graph, we do this because it makes it very easy to compare one object with another, or how the speed of one object changes over time.

Display the first slide and add to the following diagram with each step accompanied by an explanation, similar to the one that follows.

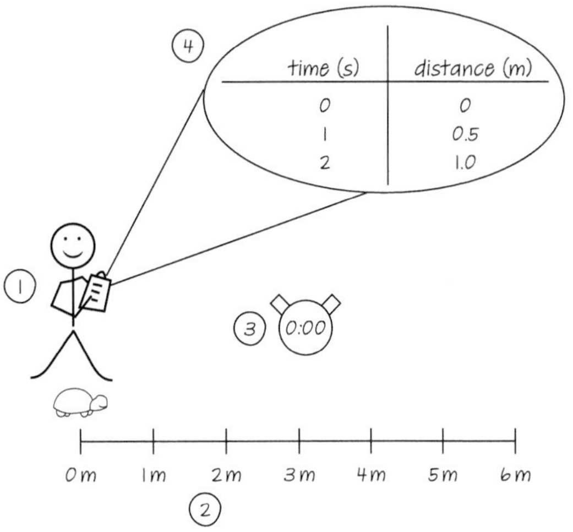

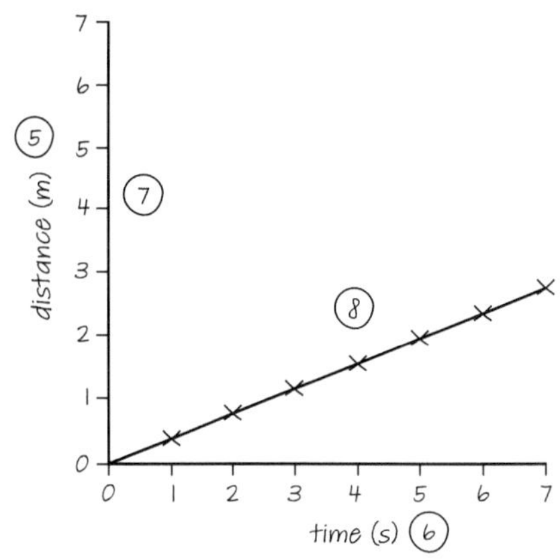

1 We have an observer – our scientist – watching the progress of a tortoise moving over time.
2 We have our distance markers to measure the distance laid out.
3 And we start our stopwatch.
4 We keep a record of our results in a table, a bit like we have done before. We keep a meticulous record of the distance over time. However, it is not particularly easy to see trends from this table, so we plot the data.
5 We put the distance on the *y*-axis.
6 We put the time on the *x*-axis.
7 We add appropriate scales to both of our axes, which fit with our data.
8 If students have covered Unit B2 already, then use the phrase *line of best fit*. If not, just say that we will draw a line to connect the points.

Display the second slide to show the graph more clearly. Explain that we call this a distance-time graph because it shows how the distance changes as time progresses. In this case, we can see exactly where the tortoise was at any moment in time.

Explain this further by adding to the diagram on the third slide, as shown below, with each step accompanied by an explanation, similar to the one that follows.

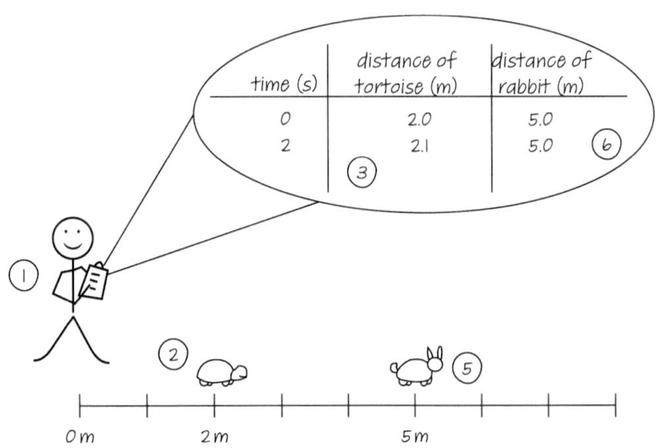

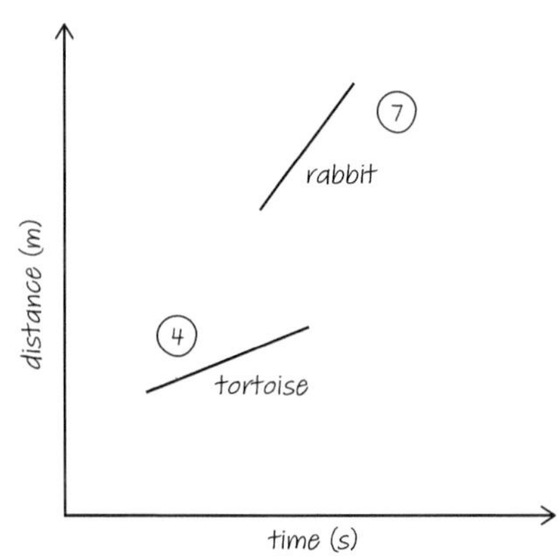

1 If we were to measure the speeds of different animal species, they might not behave themselves. They may start moving from random places and start at random times. This is not ideal, but it is something that scientists can deal with.
2 Let us say that our tortoise starts 2 m away from where we are standing. (This can be drawn as a box labelled T instead of doodling a tortoise.) This is completely fine, provided that we keep meticulous records of the time and the distance travelled by each animal.

3 So we record the time and the distance of the tortoise, even if the tortoise does not move for the first couple of seconds
4 Rather than comparing the data in the table, it is easier to see trends and compare if we plot the data as a graph. We plot the time data on the *x*-axis (the horizontal axis) and the distance data on the *y*-axis (the vertical axis).
5 But we can do this for multiple animals to compare them on the same graph. Let us say there is a rabbit that starts even further away than the tortoise, but it only starts moving a bit later.
6 This is fine, as long as we keep good records of all the data.
7 We can now plot this new set of data for the rabbit on exactly the same graph as the tortoise, so we can make a close comparison. On this graph we can see two things straight away: the rabbit started its journey later on (because the line starts further along the time axis) and the rabbit started further away than the tortoise did (because the line starts further along the distance axis).

Check for understanding

Display the *CFU* slide and, using MWBs, ask questions like the ones below.

- On which axis do we put distance on a distance-time graph?
- On which axis do we put time on a distance-time graph?
- What do we measure distance in?
- What do we measure time in?
- What goes on the *x*-axis of a distance-time graph?
- What goes on the *y*-axis of a distance-time graph?

There is no independent practice at this stage. If you think your students need some consolidation, ask them further questions of a similar style to the CFU.

Guided explanation

How do we read values on a distance-time graph?

Display the graph on the first slide. Click to reveal each step sequentially, adding labels as shown below, accompanied by student input or explanation, similar to the outline that follows.

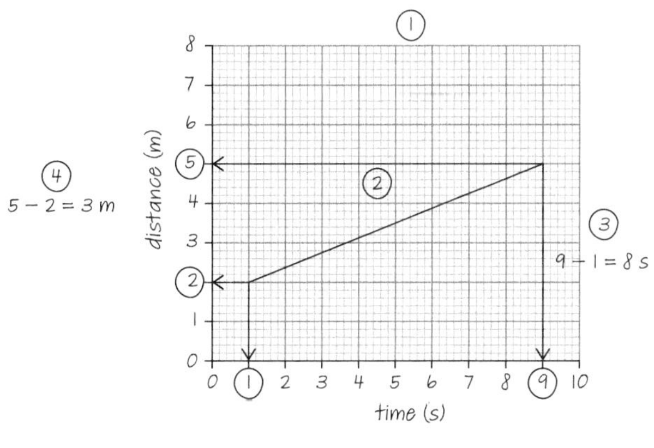

1 Let students look at the graph line and give them time think about what the data show.
2 Call their attention back to you and use MWBs to test that they can read the data correctly, e.g. by asking what the distance is at 5 seconds.
3 Show them that we work out the time that the blue object has been travelling for by subtracting the time when the line started (1 s) from the time when the line finished (9 s). Click on the slide to display this. Add your calculation 9 - 1 = 8 s to the slide.
4 Demonstrate how we work out the distance that the blue object has travelled by subtracting the distance when the line started (2 m) from the distance when the line finished (5 m). Again, click on the slide to display this and write 5 - 2 = 3 m on the slide.

Display the second slide and repeat these steps for the next example. The red object starts moving at 3 s and finishes moving at 8 s, so it is moving for a total of 5 s (8 - 3 = 5). The red object starts 4 m away from the observer and finishes 7 m away from the observer, so it has travelled a total of 3 m (7 - 4 = 3).

Manipulate the lines as necessary to create additional examples, if required.

Check for understanding

Display the first *CFU* slide and, using MWBs, ask questions like the ones below.

- At what time does the green object start moving?
- At what time does the green object stop moving?
- How long is the green object moving for?
- How far away from the observer does the green object start?
- How far away from the observer does the green object end?
- How far does the green object travel?

Repeat for the second slide and as many extra examples as are necessary (be careful with the choice of numbers, e.g. starting at 2 m and finishing at 4 m is a bad choice because 4 - 2 = 2, and this is one of the original numbers, so students could get the correct answer for the wrong reasons).

There is a blank graph on the third slide for you to add lines to, should you need to.

Independent practice review

Student Practice Book questions 79–94 (pages 98–101).

Q	Notes
85, 86	The question is phrased differently to remove some of the scaffolding, so ensure that students read these ones carefully.
91, 94	The questions have now removed almost all scaffolding, so students are finding distance travelled and time taken by themselves. Add back in the additional Qs if necessary for support.

Guided explanation

How do we calculate speed on a distance-time graph?

Display the first slide and work through the example, adding labels sequentially as shown below. Use the same types of annotation as previously to find distance travelled and the time taken. Clearly show that you cannot just read these values, but need to perform subtractions too.

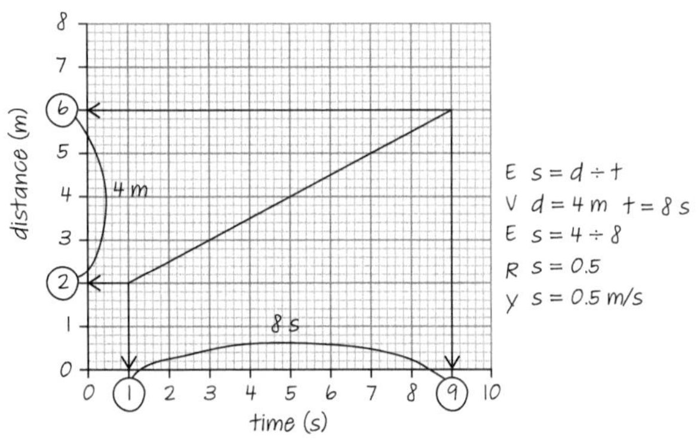

1 The blue object has travelled 4 m in 8 s.

2 Show a calculation of speed using these values and the EVERY method, using students' prior knowledge of the speed equation and how to set out calculations.

Display the second slide and work through the example, adding labels sequentially as shown below.

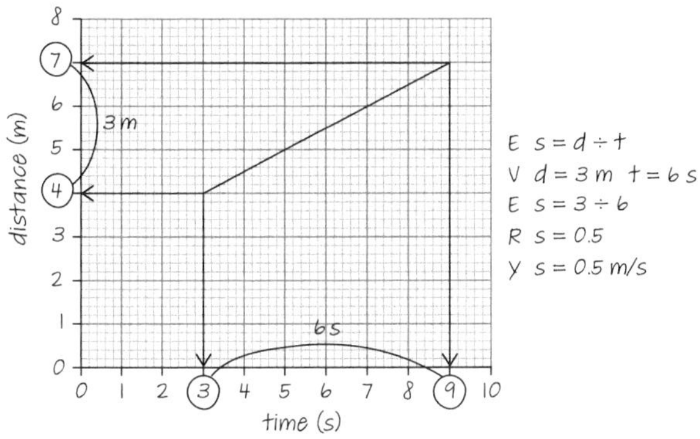

1 The red object has travelled 6 m in 3 s.

2 Show a calculation of speed using these values and the EVERY method, using students' prior knowledge of the speed equation and how to set out calculations.

Check for understanding

Display the first *CFU* slide and, using MWBs, ask questions like the ones below.

- How far does the green object travel?
- How long does it take to travel 6 m?
- If it travels 6 m in 3 s, what is its speed? *Ensure that students show their working out.*

Display the second slide and ask questions like the ones below.

- How far does the purple object travel?
- How long does it take to travel 6 m?
- If it travels 6 m in 2 s, what is its speed? *Ensure that students show their working out.*

The example on the third slide includes different *x*- and *y*-scales. This is something students should have seen in Unit B2, but given how many steps there are in interpreting a distance-time graph, we are not going to focus on different scales. This is a single example just to give students exposure to other scales and to think about how to apply their approach.

Display the third slide and ask questions like the ones below.

- How far does the orange object travel?
- How long does it take to travel 9 m?
- If it travels 9 m in 18 s, what is its speed? *Ensure that students show their working out.*

Repeat for as many extra examples as are necessary. There is a blank graph on the fourth slide for you to add lines to, should you need to.

Independent practice review

Student Practice Book questions 95–103 (pages 101–04).

Q	Notes
99e	We are using the remaining distance for the second speed, so 2000 m minus the initial 600 m, which is 1400 m.

Guided explanation

How do we compare speeds on a distance-time graph?

Display the slide and work through the example, as shown below. Add the labels sequentially, with each step accompanied by student input or explanation, similar to the outline that follows.

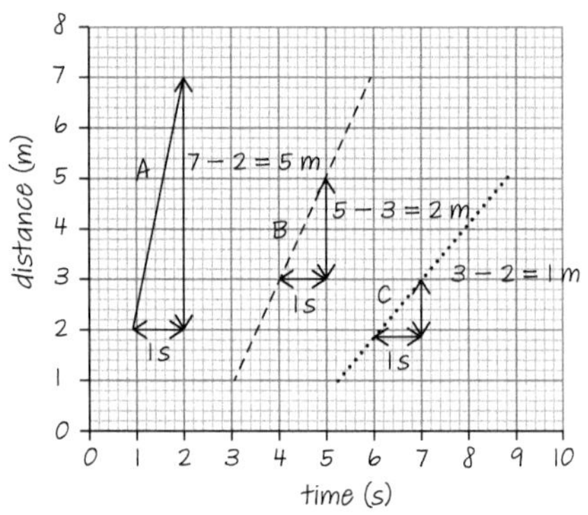

1. Tell students that we originally said graphing was useful because we could easily compare different moving objects. Here are three different objects and, just by looking at the steepness of the lines, we should be able to work out which is faster.
2. Explain that we could calculate the speed as we have already been doing, and this is perfectly fine to do. However, we could also look at how far each object travels in 1 s. Remind students that speed is simply how far something travels in 1 s of time.
3. Annotate each line to show how to take just 1 s of time and show much the distance has changed by in this time. In 1 s, object A moves 5 m, so the speed is 5 m/s. In 1 s, B moves 2 m, so the speed is 2 m/s. In 1 s, C moves 1 m, so the speed is 1 m/s.
4. Explain that the fastest object, A, went 5 m/s and because of this its line has to be the steepest so the distance can increase by the most in each second. The slowest object, C, went 1 m/s and because of this its line is the shallowest since it travelled the least distance in each second.
5. Ensure that students understand that the steeper the line, the faster the object, and the shallower the line, the slower the object.

Check for understanding

Display the *CFU* slide and, using MWBs, ask questions like the ones below.

- Which object is the fastest?
- How do you know?
- Which object is the slowest?
- How do you know?
- How do you know blue is faster than green?
- How do you know blue is slower than red?

Independent practice review

Student Practice Book questions 104–09 (pages 104–05).

Q	Notes
107	Ensure that students make reference to time in their answer.

Guided explanation

What are some special cases on a distance-time graph?

Display the slide, which shows a graph for two stationary objects, accompanied by an explanation, similar to the one below.

1. We can also show stationary objects on a distance-time graph. Stationary objects are objects that do not move, so their distance does not change at all over time. This means that a stationary object is represented by a horizontal line on a distance-time graph.
2. The blue object is at a distance of 4 m. This does not change between 2 s and 8 s, so the object is stationary for 6 s (8 - 2 = 6).
3. The green object is at a distance of 2 m. This does not change between 6 s and 10 s, so the object is stationary for 4 s (10 - 6 = 4).

Lots of students will say that the line for a stationary object is *straight*. But *all* the lines we have looked at have been straight. We need to say that stationary objects have a *horizontal line* on a distance-time graph, and that moving objects have a *slope* on a distance-time graph.

Check for understanding

Display the *CFU* slide and, using MWBs, ask questions like the ones below.

- Which object is stationary?
- How do you know? *Do not accept 'straight' because all the lines we have looked at are straight.*
- Which objects are moving?
- How do you know?
- Which object is fastest?
- How do you know?
- Which object travels a total of 2 m?
- How do you know?
- Ask students to calculate speeds of certain objects.

Independent practice review

Student Practice Book questions 110–15 (pages 106–07).

Q	Notes
110	'Straight line' is a common wrong answer. All of the lines we have looked at are straight because they can be drawn with a ruler. A stationary object has a line that does not increase on the *y*-axis, and we call this horizontal
111	To be consistent with the Knowledge Book, we are using the word 'slope' to denote something with a gradient.

Guided explanation

What do more realistic journeys look like?

Explain that, during a typical journey, we do not just travel at one speed. Therefore, a distance-time graph can show how the speed changes over time.

1. Display the graph on the first slide, without the *x* and *y* scales. Ask students to identify the fastest part, the stationary part and the part where the speed is lowest but it is still in motion.
2. Click to add the scales. Isolate individual parts and show students how to calculate the speed for each part using the EVERY method. They should be familiar with this by now, and you can start bringing students into your explanation via questioning or MWBs.
3. Repeat the process for the example on the second slide, but with more input from students.

Check for understanding

Display the *CFU* slide and, using MWBs, ask questions like the ones below.

- In which section is the object stationary?
- How do you know? *Do not accept 'straight'.*
- In which section is the object fastest?
- How do you know?
- How far does the object travel between 3 s and 5 s (the green section)?
- What is the speed in the green section? *Ensure that students show working out using EVERY.*
- How long does it take the object to go from 3 m to 6 m (the red section)?
- What is the speed in the red section? *Ensure that students show working out using EVERY.*

Independent practice review

Student Practice Book questions 116–20 (pages 107–08).

Q	Notes
116a	We are focusing on non-zero speeds, so there are three.
116b	Ensure that students calculate the difference between the start and end of the stationary period.

P2.3 Relative motion

Students' learning objective: *I am learning about relative motion so that I can compare the motion of one object with that of another.*

We focus purely on one-dimensional examples when looking at relative motion, so speeds either add up or subtract from one another. We will try to get students to see this through the lens of how fast objects will 'feel'.

Knowledge check

After working through the *General science quiz*, display the *Knowledge check* slide and, using MWBs, deliver the questions one by one, reteaching concepts as necessary.

	Question	Answer	Notes
1	What does 1 m/s mean?	One metre travelled every second	Do not accept 'how fast' – push for more detail.
2	If I am travelling at 5 m/s, how far do I go in 1 s?	5 m	
3	What does stationary mean?	Not moving – speed is zero	

	Question	Answer	Notes
4	A bird flies for 8 s and travels a distance of 56 m. What is the speed of the bird?	7 m/s	Ensure that students show working out using EVERY.
5	What does relative mean?	Compared with	Students could give lots of answers here, but show how familial relatives and the like all link back to the concept of a link between two things.

Guided explanation

What if the observer is not stationary?

Introduce the topic with the scenario of a student who sees their friend running towards them. Draw the diagram below sequentially, with each step accompanied by an explanation, similar to the one that follows.

1 Imagine that you are stationary and your friend is running towards you. It takes them 10 s to get to you.

2 Now imagine this again, but as soon as you see them, you run towards them. You reach them much quicker.

3 It *feels* much faster when you are both moving towards one another. The speed between the two of you must be greater if you meet in a quicker time. (Emphasise the *feeling* at this point.)

Move on to a scenario of a student standing by the side of a road. Draw the diagram below sequentially, with each step accompanied by an explanation, similar to the one that follows.

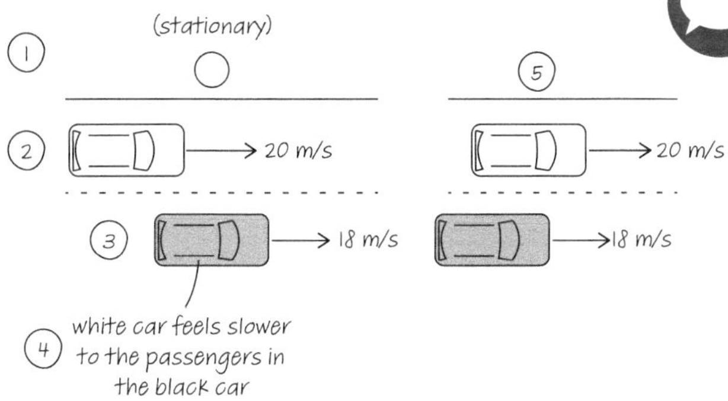

1 Sometimes things can feel like they are moving slower than they actually are. Imagine standing by a road where the cars are travelling really quickly.

2 You see a white car zoom by and to you it feels like it is really fast. It is travelling at 20 m/s, so it is definitely fast. It passes you by in less than a second. Sometimes however, to other people, this same car can *feel* like it is moving much slower.

3 Let us say there is another car on the road, a black car. It is ahead of the white car so passes you by first. It whizzes past you quickly as you stand on the side of the road. You probably cannot tell the difference between the speeds of the black and white cars.

4 To a passenger in the black car, the faster white car is catching up with them, but it is catching up *really slowly*.

5 It will eventually overtake them, but it takes a long time. And because it takes a long time to catch up and to overtake, the white car *feels much slower* to the people in the black car than it did to you standing at the side of the road. The speed between the two cars must be slower if it has taken so long for the overtake to happen.

Explain relative motion further by drawing the diagram below sequentially, with each step accompanied by an explanation, similar to the one that follows.

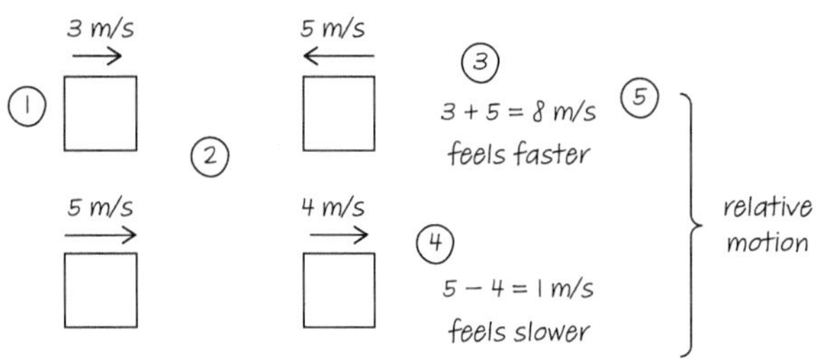

1 Here are two sets of moving objects – one set going in opposite directions and one in the same direction. In each case we are looking at the *relationship* between the two moving objects.
2 Add numbers to the arrows.
3 In the first example, we say they have a relative motion of 3 + 5 = 8 m/s. We are adding them together because it has to *feel faster* when they are moving towards one another.
4 In the second example, we say they have a relative motion of 5 - 4 = 1 m/s. We are subtracting them because it has to *feel slower.*
5 These numbers are the *relative motions*.

Emphasise to students that they can use the worked examples in Topic P2.3 of the Knowledge Book to help them if necessary.

Check for understanding

Display the first two *CFU* slides, for objects moving in opposite directions. Using MWBs, ask questions like the ones below.

- What is the relative speed between the blue and orange objects?
- Why do you add them? *Accept 'towards one another' or 'has to feel faster'.*
- What is the relative speed between the grey and green objects?
- How do you know?

Display the next two slides, for objects moving in the same direction. Ask questions like the ones below.

- What is the relative speed between the blue and orange objects?
- Why do you subtract them? *Accept 'away from one another' or 'has to feel slower'.*
- What is the relative speed between the grey and green objects?
- How do you know?

Display the final slide on relative speed and ask questions like the ones below.

- Which pairs will feel faster?
- How do you know?
- Which pair has a relative speed of 1 m/s? *Students may say A if they are doing it wrong.*
- Which pair has a relative speed of 8 m/s? *Students may say B if they are doing it wrong.*

Independent practice review

Student Practice Book questions 121–32 (pages 109–11).

Q	Notes
130, 131	Both of these explore the comparison between relative speed and absolute speed. We have not introduced the concept of absolute speed, so you can just use the phrase 'actual speed' when working through solutions.

P3 Forces

P3.1 Basic forces and diagrams

Students' learning objective: *I am learning about forces so I can identify the forces acting on a range of examples.*

Students should have a good understanding of forces from KS2. They should be well aware of friction, gravity and magnetism. This unit starts with a simple review of forces and builds towards using moments to explain simple machines.

Knowledge check

After working through the *General science quiz*, display the *Knowledge check* slide. The questions are best delivered one by one on mini-whiteboards (MWBs). If students do not know the answer to any of them, you should reteach the relevant concept. Answers are not on the slides because it is best to go over them together as a class..

	Question	Answer	Notes
1	What is a force?	A push or pull	These should be known from the KS2 National Curriculum.
2	What do we call the force between two magnets?	Magnetism	
3	What do we call the force that occurs when two objects rub together?	Friction	

Guided explanation

What is a force?

Introduce the topic by drawing the diagram below sequentially, with each step accompanied by an explanation, similar to the one that follows.

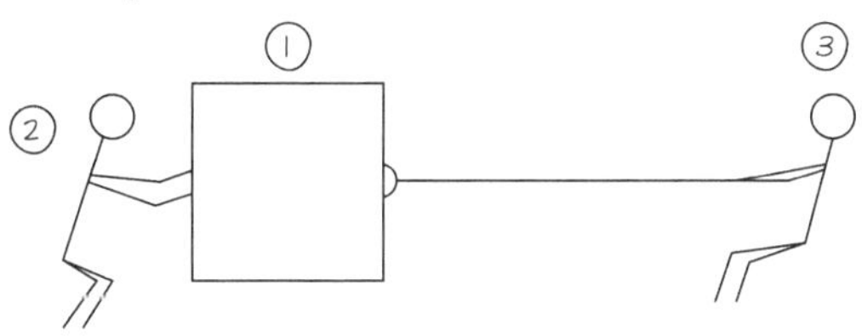

1. Start by drawing a large box.
2. Add a person giving it a push. Tell students that the person is trying to use a force to change the box's movement.
3. Then add a person trying to pull the box. Explain how the pull is also trying to change the movement of the box. Both the push and the pull are about trying to change the box's movement, and we call them forces.

Explain that there are many different forces in the universe, but essentially they all work as either pushes or pulls. We can simply say a force is a push or pull at this time. Some forces only push *or* pull. Weight is a force (due to gravity) that pulls objects towards the ground. Other forces can push *and* pull. Magnetism can pull two opposite poles together or push away two like poles (you might want to demonstrate this).

Check for understanding

Display the *CFU* slide and, using MWBs, ask questions like the ones below. Tailor and adapt your questions to the class in front of you.

- What are all forces?
- Is magnetism a push, pull or both?
- Is gravity a push, pull or both?

Independent practice review

Students practise independently by answering questions 1–3 from their Practice Books (page 121). Full answers can be found on the *IP answers* slides or in the answer documents on Boost. Notes are only included for questions, where useful.

Q	Notes
3	Students can say the answer is wrong as long as they justify with the correct reasoning.

Guided explanation

How do we show the direction and size of a force?

Start by telling students that we know that forces are either pushes or pulls and that they come in different sizes, but we have a problem. We cannot see forces. Point out that we can easily see the effect of forces, but we cannot see the forces themselves. This makes it hard to show when a force is acting.

How do we show the direction of a force?

Draw the person with the box again, as shown below, constructing it sequentially, with each step accompanied by an explanation, similar to the one that follows.

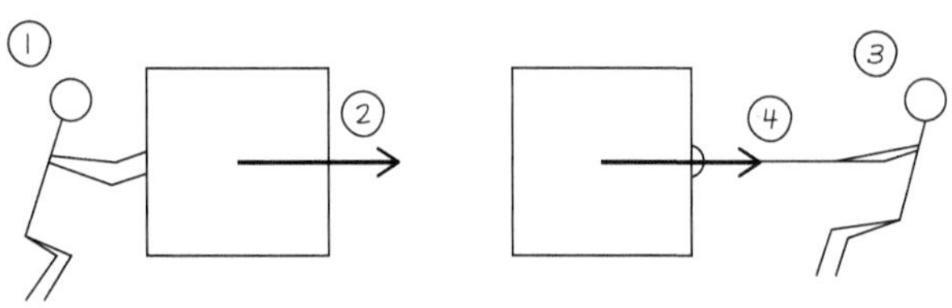

1 (Draw the person pushing the box.) There is a force pushing to the right, but it cannot be seen.
2 (Add an arrow pointing in the direction of the pushing force.) Now we can see the force.
3 (Draw the person pulling the box.)
4 (Add the force arrow to show the force.) Compare the two arrows. (Allow time for student responses.) Both attempts are trying to make the box move to the right and the forces look identical.

Note that in *Springboard*, to ensure consistency, we will always draw force arrows from the centre of mass, but students do not need to worry too much about their own precision here.

Reiterate that we use arrows to show the directions of forces on an object and that forces can act in all directions. Draw some stick people lifting or pulling boxes, as shown below. Or, if preferred, you could physically demonstrate forces acting in different directions.

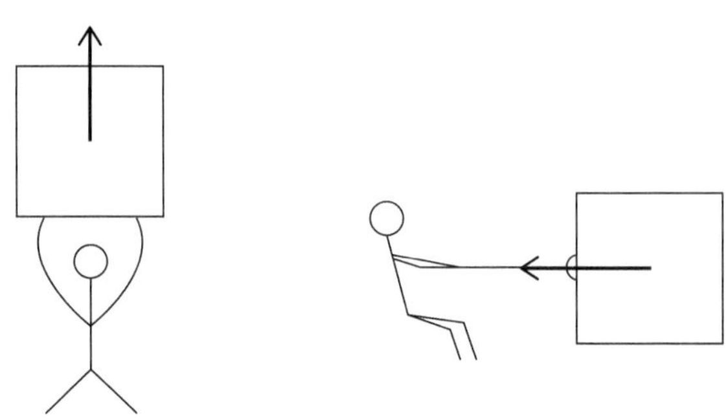

How do we show the size of a force?

Using an explanation similar to the one below, tell students that direction is not the only thing we need to show about forces. We need to also show how large forces are.

1 Not all forces are the same strength. We measure how hard the push or pull of a force is and give it a number.
2 In science, all measured numbers have a unit, so we know what they are measuring. The unit of force is the newton (N), named after Isaac Newton. A larger force has more newtons than a smaller force.
3 We measure the size of a force with a piece of equipment called a *newtonmeter*.

Display the image of a newtonmeter on the second slide and, ideally, also demonstrate using a newtonmeter to measure the force pulling on something like a stool or block.

Illustrate different-size forces using the person with the box again accompanied by an explanation, similar to the one that follows.

1 (Draw a small arrow to represent a small force.) When we draw arrows for forces, we use the direction of the arrow to show which way the force is acting on the object and the length of the arrow to show how large the force is. We do not use the thickness of the arrow.
2 (Draw a longer but not wider arrow to represent a larger force. Make this difference clear to students.) A larger or thicker arrow is not correct. The arrow should be *longer* if the force is greater.
3 (Add values to the force arrows.) If we do not draw a force diagram, then we use numbers to show the size of each force. Add values, e.g. 5 N and 25 N, to each arrow to illustrate.

Check for understanding

Display the *CFU* slide and, using MWBs, ask questions like the ones below.

- Which force is larger, A or B?
- Which force will cause the box to move right?
- In what way are the two forces similar and in what way are they different?
- What do we use to show the direction of a force acting on an object?
- What feature of the arrow shows how large a force is acting on an object?
- Draw a box that has a large upwards force pulling on it.
- Draw a box that has a small leftwards pull acting on it.
- Draw a box which has identical forces acting on the left and on the right of it.

Independent practice review

Student Practice Book questions 4–12 (pages 112–13).

Q	Notes
6	Students often draw longer arrows thicker, but this should be regularly corrected.
11, 12	This question refers back to prior content, so students should be encouraged to look at Unit P1 of the Knowledge Book for support, if needed.

P3.2 Naming and categorising forces

Students' learning objective: *I am learning how to name forces so I can explain how forces interact.*

Students should be comfortable drawing forces acting on an object. We will now need to introduce the ten main forces and help students to categorise them. Students will have formal and informal knowledge of the different forces from prior education and personal experiences, but it is best to avoid the temptation to assume they know it all because they will have different gaps in their prior knowledge.

You will notice that tension is covered in this topic but is not included in the Knowledge Book. This is to ensure a focus for Year 7 on the most important and clearly defined forces. Electrostatic forces have also been omitted from this topic to reduce the number of forces that students are introduced at this stage. Unit P8 later in the course will focus on electrostatics and its importance.

Knowledge check

After working through the *General science quiz*, display the *Knowledge check* slide and, using MWBs, deliver the questions one by one, reteaching concepts as necessary.

	Question	Answer	Notes
1	What is a force?	Push or pull	
2	What are forces measured in?	Newtons	Accept N.
3	What device is used to measure forces?	Newtonmeter	
4	What do we use to show forces acting on an object in diagrams?	Arrows	
5	What do we change to show the different sizes of forces acting on objects in diagrams?	The length of the arrow	Make sure that students do not refer to the 'size of the arrow'.

Guided explanation

What are the main types of force?

This explanation covers the ten main forces – it will introduce the forces and some contexts.

On the first slide, write *Ten forces*. After each force is introduced, go back to this list and add the name of the force underneath. This will help show students what you have done so far, as well as provide a consistent thread through the explanation.

Explain each of the forces using stepped explanations, similar to the ones below.

You will need a magnet and an iron nail to demonstrate magnetism.

Applied force

1. *Applied force* is the name we use for any time a person applies a simple push or pull force.
2. Therefore, if we push something (demonstrate this), it is an applied push force.
3. And any time we pull something (demonstrate this), it is an applied pull force.
4. These forces do not have to be horizontal. We can push or pull things in any direction (demonstrate this).

Note that this force is listed and named here in order to make examples simple to understand. It does not have as strict a definition as some of the other forces, but students will be able to get the hang of it via examples.

Weight

On the second slide, draw an apple hanging from a branch, as shown below.

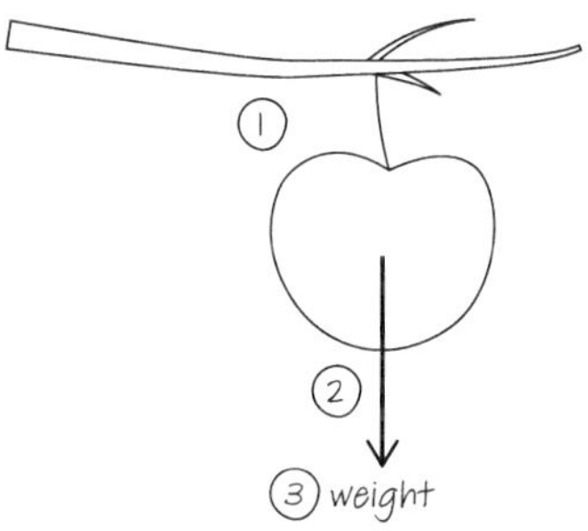

1 (Ask students) What will happen to the apple if the stem breaks?
2 The apple falls because it is being pulled towards the Earth.
3 The force that pulls the apple to the Earth is called *weight*. Weight is the force that occurs when gravity pulls on an object with mass.

Thrust

1 Engines provide an applied push force that make objects move.
2 We call this force a *thrust*.

Friction

Rub your hands together. Say that when two objects move past each other, they rub. Then draw a person pushing a box, as shown below.

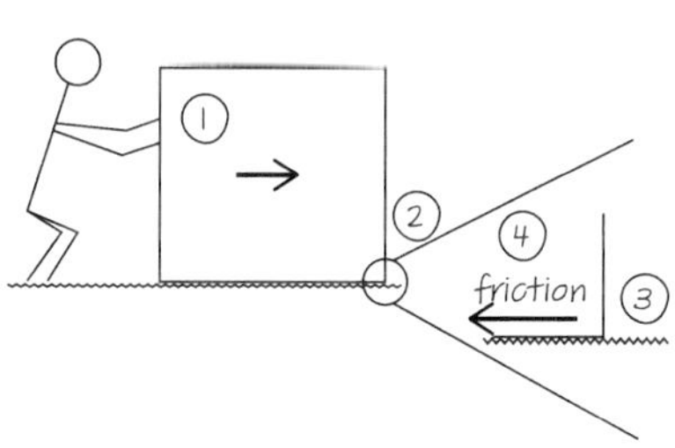

1 The push force is pushing to the right. The box will rub along the floor as it moves. This box will not move easily. Why is that?
2 If we zoom in, we can see the reason.
3 The surfaces are not smooth.
4 As the box moves, it gets caught on the tiny bumps and they provide a small force against the motion. All these tiny forces added together are called *friction*.

Air resistance and water resistance

1 These two forces are very similar, so we have grouped them together. They work in the same way but in different situations.
2 When you move your hand through the air, you can feel a force of air (demonstrate this).
3 When you move your hand through water, you feel an even greater force. These are both types of friction because they push against, or resist, your movement.
4 The friction is caused by your hand trying to push the air or water particles out of the way and hitting them. We call it *air resistance* or *water resistance*.

Students often find air resistance hard to remember because they cannot see the particles, so going back to the demonstration of the hand moving through the air might provide a useful cue if they get stuck later.

When you return to your list on the board to add friction, air resistance and water resistance, point out their similarity in that they are all forces that oppose movement.

Upthrust

Draw a rubber duck on water, as shown below.

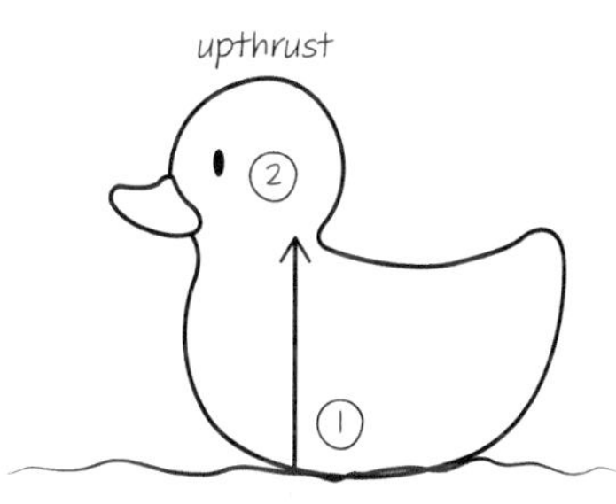

1 (Ask students) Why does it not sink? Weight must be pulling it down.
2 The water must be pushing it up. We call this force *upthrust*.

Lift

Draw a plane, as shown below.

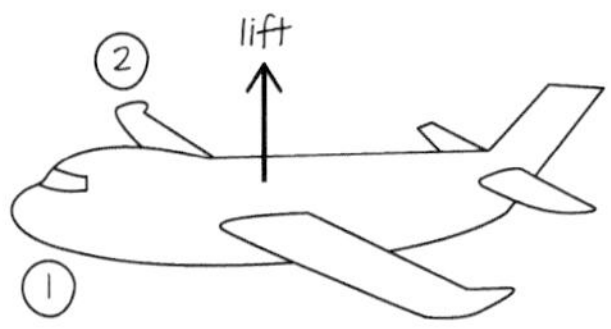

1 (Ask students) Why does the plane not fall to Earth?
2 The force that keeps flying objects from falling is called *lift*. It is a bit like upthrust, but it is not the same. A lot of students accidentally call it upthrust, but it is different.

When you return to your list to add upthrust and lift, point out their similarity as forces that stop things falling down.

Normal contact

Draw a book on a table, as shown below.

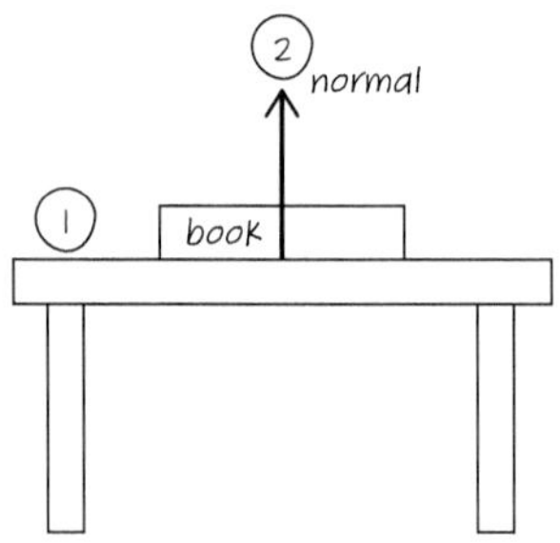

1 If we put a book on a table, the weight of the book is pulling down. However, the book does not fall. This is because the table is stopping it. It the table were made of paper it would not stop the book and it would collapse.
2 There must be a force-resisting force preventing this. This force is called the *normal contact* force. It is called *normal* because that is the maths word for ninety degrees, and the force acts perpendicular to the surface of the object. It is called *contact* because the force is not there until the objects touch.

When you add normal contact force to your list, tell students it is similar to upthrust and lift in that it stops objects falling, but in this case it stops them falling through solids.

Magnetism

Using a magnet and an iron nail (under a visualiser), demonstrate magnetism.

1 The magnet attracts the iron. I can feel them pull together.
2 This is the force of *magnetism*. We will revisit this a little later in more detail.

You should now have finished your list of ten types of force. Display the list and emphasise that in every case we discuss, we are going to be using one of these forces.

Check for understanding

Display the *CFU* slide and, using MWBs, ask questions like the ones below. It is advised that you display your list of forces and explicitly show students how to use it to answer the questions.

- What force prevents planes falling to Earth?
- What force keeps a boat floating?
- What force opposes the motion of a box being pushed along a floor? *Ensure that students are clear on what the word 'opposes' means.*
- What force opposes the motion of a ball through the air?
- What force stops a shoe sinking into the mud? *Normal contact and upthrust are both legitimate, depending on the fluidity of the mud. The closer it is to a puddle, the closer it is to upthrust.*
- What force pulls on a child sitting on a tree branch and what force stops the child falling from the branch?
- What force could I use to lift iron filings from a desk?
- What force opposes the motion of a shark swimming through the ocean?
- An object is falling and a student says that 'gravity force is acting on it'. Explain why the student is wrong.

- What force pulls an icicle off a roof?
- What force causes me to graze my knee when I fall over?
- Why is it easier to pull a box on a wood floor than on a carpet?
- Why is there no air resistance on the Moon?
- A fish is swimming straight ahead. A student says 'upthrust is slowing it down'. Explain why the student is wrong.

Independent practice review

Student Practice Book questions 13–20 (pages 113–14).

Q	Notes
20	This question links P3 back to P2, so students should be encouraged to look back at Unit P2 of the Knowledge Book if needed.

Guided explanation

What are interactions?

The idea of interactions between two objects that create pairs of forces is complex, and so it is suggested that you use the information from the students' performance in prior areas of the unit to adjust the depth of your explanation and possibly simplify it if required for some students. The most important aspect of the explanation is to build an awareness of the types of force students will come across in their science education.

Explain that what we have seen so far is that forces *always act between two objects.*

1 For example, a magnet cannot be attracted or repelled without there being a second magnet.
2 We call this an *interaction.* The two objects *interact* with one another.
3 The forces between the objects always come in pairs. Each object experiences a force from the other and the forces are always in the opposite direction.
4 Illustrate this by working through the following examples, using stepped explanations, similar to the ones below. You will need two bar magnets to demonstrate magnetism.

Magnetism

Take two bar magnets. Place them so that they are visible to the class (under a visualiser). Hold them a distance apart, oriented so they will attract. Draw a diagram as shown below, (possibly placing the magnets on top of it).

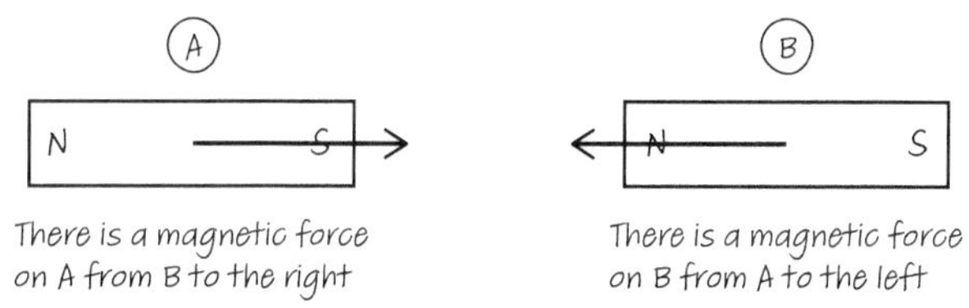

1 Magnets A and B will attract because the poles are opposites of one another. Therefore, they will move together. (Let the magnets go, so the students see them move together.)
2 When we say *attract*, what we mean is that there is a force *on each of the magnets from the other.* (Add arrows to show the direction of the force that each magnet experienced.)
3 There is a magnetic force *on* B *from* A to the left. But there is *also* a magnetic force *on* A *from* B to the right.

We will use this phrasing every time we look at forces – there is a (name of force) *on* (object 1) *from* (object 2) (state direction).

Repeat the above process, but with magnets that repel. Vary the direction so that it is not always left and right, as shown below. Be highly specific with the phrasing.

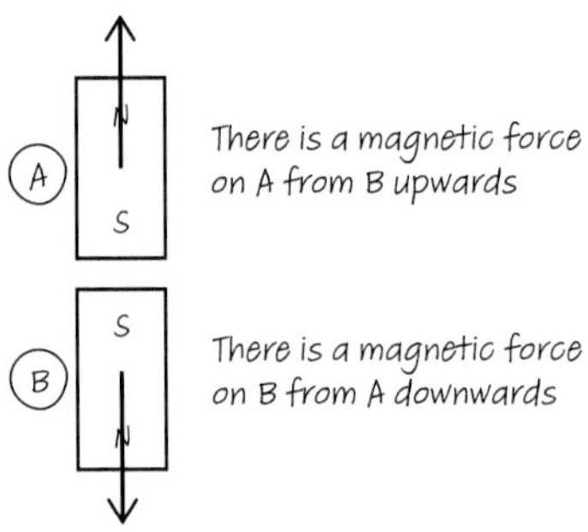

Tension

Draw the diagram below sequentially, with each step accompanied by an explanation, similar to the one that follows.

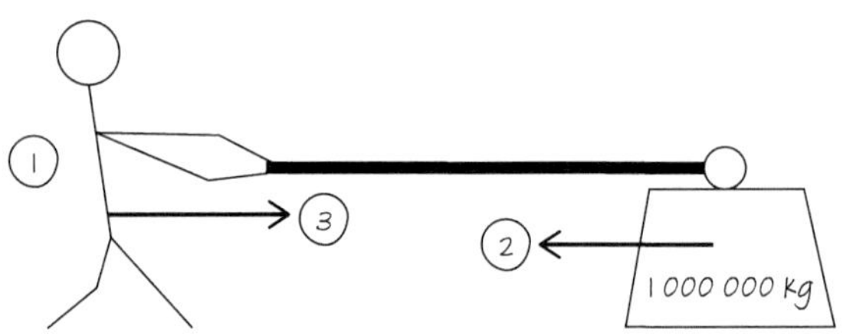

1 Imagine pulling something immovable. You can *feel* the object pulling against you.
2 (Add the arrow) There is a tension force *on* the immovable object *from* the person to the left.
3 (Add the arrow) But there is *also* a tension force *on* the person *from* the immovable object to the right.

Normal contact

Draw a bat and ball sequentially, as shown below, with each step accompanied by an explanation, similar to the one that follows.

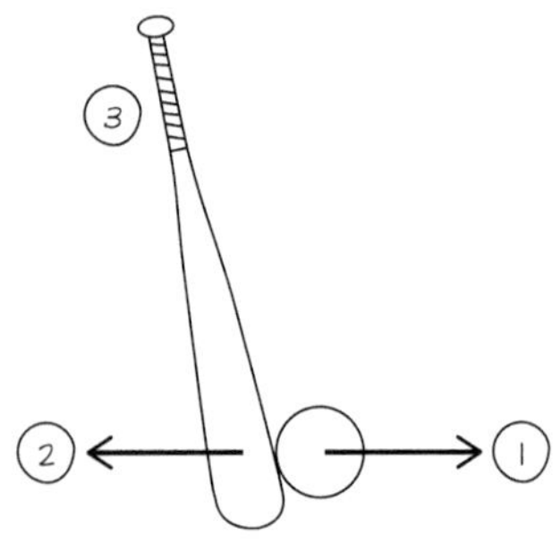

1 There is a normal contact force *on* the ball *from* the bat to the right.
2 There is a normal contact force *on* the bat *from* the ball to the left.
3 If you are holding the bat, you can *feel* this second force as the recoil of the bat. You know you have hit the ball because you feel it. You feel the second part of the interaction and the ball feels the first.

Weight

Explain that the idea of force interactions is true for all forces that act between two objects. Sometimes the interaction may be a bit counterintuitive, but it is true. Illustrate this by drawing a person above the surface of the Earth sequentially, as shown below, with each step accompanied by an explanation, similar to the one that follows.

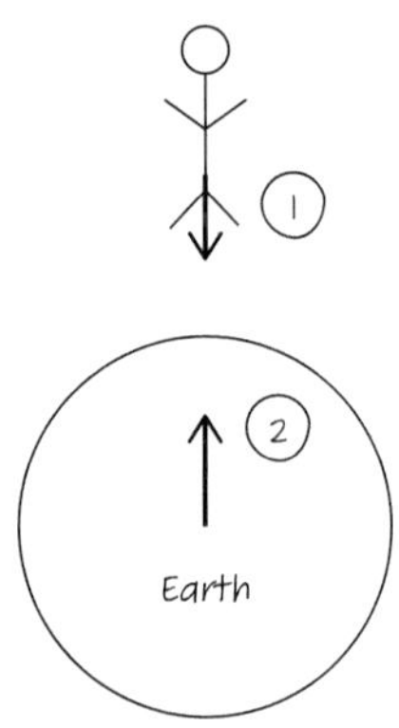

1 (Draw an arrow pulling the person down.) There's a force of weight *on* the person *from* the Earth pulling downwards.
2 But as the person and the Earth are interacting, there *must* be a second force (add this arrow). That force is a force of weight *on* the Earth *from* the person pulling upwards. This might not *feel* as obvious. But when the Earth pulls you down, you pull the Earth up too. The reason you do not notice this is because you are much smaller than the Earth, so the force is not noticeable.

Students can find the idea of interactions difficult, so for the majority of the topic we will consider the forces that act on specific objects and not draw interactions. This should allow all students to gain a valuable understanding of forces.

Check for understanding

Using MWBs, ask questions like the ones below.

- How many objects do you need for an interaction?
- If I push a wall right, which way does the wall push me? *You can demonstrate this by physically pushing a wall, and noting that you move backwards*
- If I have one magnet by itself, can there be a force?
- Display the *CFU* slide and draw two magnets. Tell the students they are repelling. Ask them to draw the pair of arrows in the interaction. (*Repeat with magnets attracting*)
- Draw a person at the Equator of the Earth. Ask them to draw the pair of arrows in the interaction. *Make sure their gravity arrow points towards the centre of the Earth.*
- What force works against moving objects?
- What force prevents a plane from falling to Earth? What are the two objects involved in the interaction? *In this case, the first object is the plane, and the second is 'the air', which is a collection of all the air particles under the plane.*
- What force prevents a swimming float sinking? What are the two objects involved in the interaction? *As above, here we have 'the float' and 'the water'.*

Independent practice review

Student Practice Book questions 21–28 (page 115).

Q	Notes
26	When describing interactions, ensure that students keep to the correct format for answering, so it is easy to check its accuracy.

Guided explanation

How do we categorise forces?

Draw the diagram below sequentially, accompanied by student input or explanation, similar to the outline that follows.

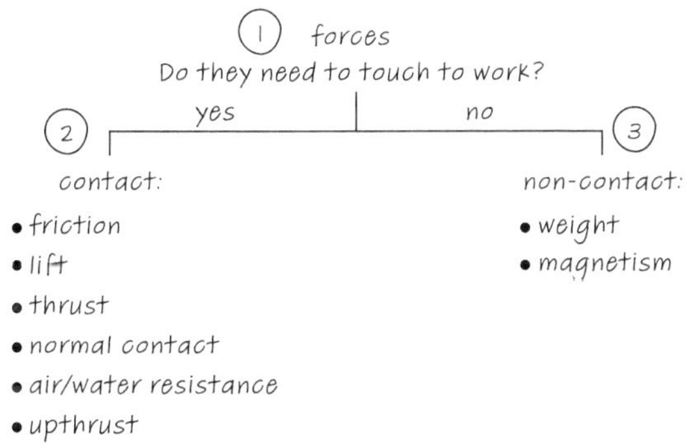

1 Explain that in order to categorise forces, we need to ask ourselves the question: *Does the force need to touch an object to interact?*

2 If the answer is *yes*, then it is a *contact force*. Talk through the following examples with students:
   - Friction – when two surfaces rub together (rub your hands to demonstrate).
   - Lift – when a plane flies as its wings move through the air.
   - Thrust – the push from engines.
   - Normal contact force – the name for the force I felt resist my push on the box in the last topic. It is the feeling of the desk pushing back as I lean on it. (Lean on a desk.)

- Air and water resistance – these need particles of air or water to bump into objects as they move. This is why objects are often made streamlined to help them move faster.
- Upthrust – the force of water pushing up, which stops things sinking if it is equal to the weight.

3 If the answer to the question is *no*, then the force is a *non-contact force*. Talk through the following examples:
- Weight – the pull of gravity on an object.
- Magnetism – the attraction or repulsion of magnets and magnetic materials.

Make sure students are aware that there are many more names for contact forces than for non-contact forces. It is worth remembering that most forces are contact forces. Students often think that air resistance is a non-contact force because the particles are too small to see, but this is wrong. Without the air particles there would be no air resistance. This is what happens in space.

Check for understanding

Display the *CFU* slide and, using MWBs, ask questions like the ones below.

- What are the two categories of forces?
- Which type of force only works when objects are touching?
- Which type of force is weight?
- Which type of force is upthrust?
- Which type of force is air resistance?
- Why do some students incorrectly think air resistance is a non-contact force?

Independent practice review

Student Practice Book questions 29–45 (pages 115–17).

Q	Notes
30	Students might include the electrostatic force here, but it is not required at this stage.
33–39	Students must provide the name of the force and categorise it as contact or non-contact to secure the mark.

P3.3 Stretching and squashing forces

Students' learning objective: *I am learning about stretching and squashing forces so I can explain why objects change shape.*

This topic will lay the groundwork for the introduction to Hooke's law. Students will have lots of first-hand experience of stretching and squashing, which can be drawn on, and having a slinky, spring or plasticine may enhance your explanations. This topic also incorporates the conversion of millimetres to metres. This is in the knowledge check, so you can determine if the class is proficient enough before deciding to teach it.

The National Curriculum introduces the word *deformation*, but this course uses *squashing* and *stretching*. You may choose to add *deformation* to your explanations if you think it is appropriate.

Knowledge check

After working through the *General science quiz*, display the *Knowledge check* slide and, using MWBs, deliver the questions one by one, reteaching concepts as necessary.

	Question	Answer	Notes
1	What do we call it when two objects act on each other using forces?	Interaction	
2	What are the two categories of forces?	Contact and non-contact	
3	How long is 33 mm in metres?	0.033 m	Make sure that students can show their working.

Guided explanation

What are squashing and stretching?

Introduce the topic by saying that we all use the words stretching and squashing and asking students what it actually means when something is squashed or stretched.

What is squashing?

Draw the diagram below sequentially, with each step accompanied by an explanation, similar to the one that follows.

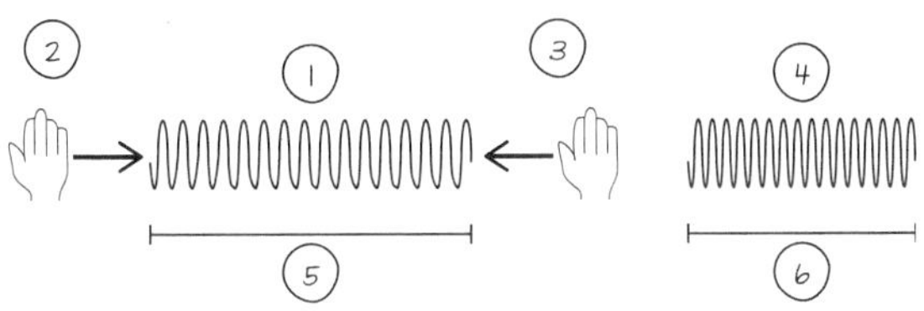

1. Let us look at an example of a spring.
2. If we add an applied force on one side, the spring will begin to move. In this case, it will move right.
3. If we add a second applied force in the opposite direction, the spring cannot move any more.
4. The spring begins to get squashed. It becomes compressed.
5. Look at how long the spring was at the start.
6. Compare this to how long it is now. It is shorter. When we squash objects, they become shorter. Another word for this shortening is *compression*.

What is stretching?

Draw the diagram below sequentially, with each step accompanied by an explanation, similar to the one that follows.

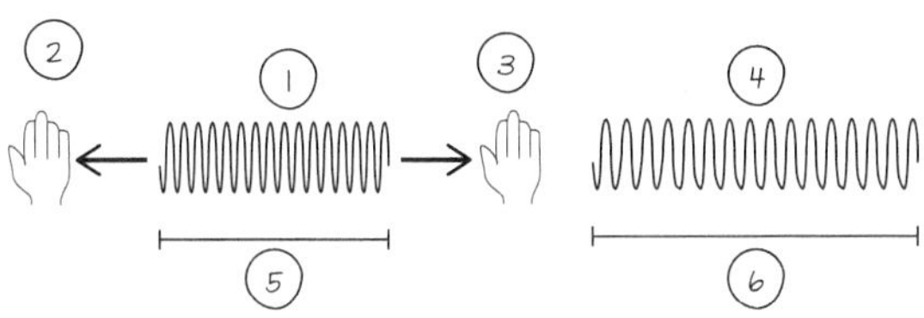

1. What about if we do the opposite to the same spring?
2. This time, we use an applied pull force on the left.
3. The spring will move to the left unless a second applied pull force is added to the right.
4. This causes the spring to stretch. It gets longer.
5. The original spring was much shorter than the stretched spring.
6. The length the spring has increased when stretched is called the *extension*.

Check for understanding

Display the *CFU* slide and, using MWBs, ask questions like the ones below.

- What do we call it when an object's length is reduced?
- Do we use pushes or pull forces to squash an object?

- When I use two pull forces in opposite directions what happens to the object?
- If I pull on one side of a spring and push on the other side of the spring what will happen to it? *Accept answers that focus on it not squashing or stretching, because motion has not been covered in detail.*
- My object's length has contracted, has it been squashed or stretched?
- Draw a block of plasticine and add two arrows to show the forces needed to extend its length. Ask whether the plasticine will extend or compress.

Independent practice review

Student Practice Book questions 46–49 (page 117).

Q	Notes
49	Make sure the students refer to the force and the direction.

Guided explanation

How do we measure changes in objects?

Measuring the extension of a spring is a vital skill. If not done correctly, it often leads to errors when investigating Hooke's law. Explicitly teaching it and practising it now will enable all students to complete the practical later in the unit error-free.

Explain the process of measuring change to students by drawing a diagram to illustrate the compression of a block of plasticine, as shown below. Draw it sequentially, with each step accompanied by an explanation, similar to the one that follows.

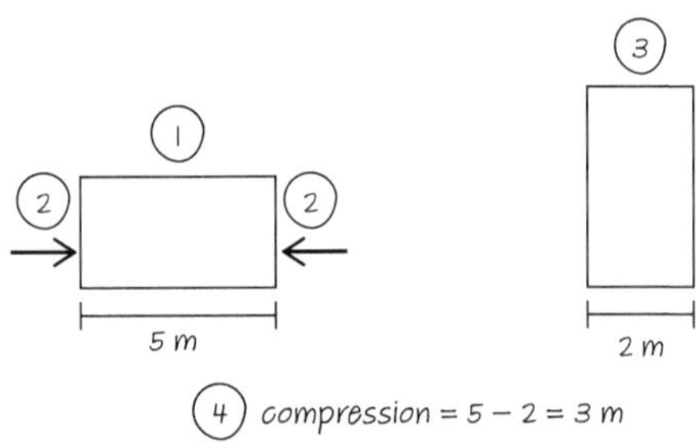

1. (Draw a large block of plasticine.) This block of plasticine is 5 m wide. It is really big. (While the number is unrealistic, the importance of ease of use while enforcing standard units makes it appropriate.)
2. (Draw a push on the left and an equal push on the right.) The block gets squashed, or compressed.
3. The block is now only 2 m wide. How much has it been compressed by?
4. To work out the compression, we take the original length and subtract the new length. In this case it has a compression of 3 m.

Repeat the process to show the extension of a block of plasticine, as shown below.

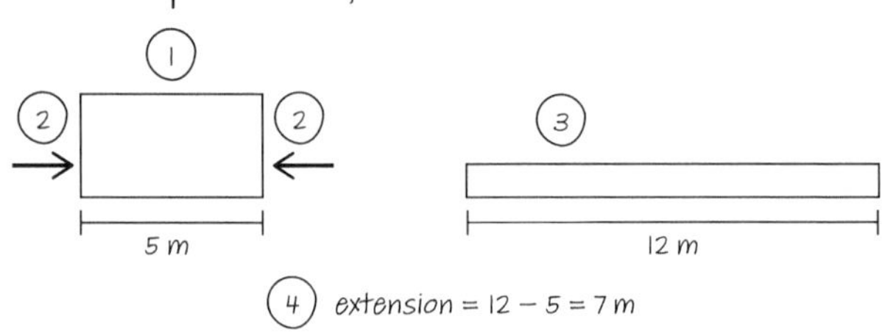

1. (Draw the same block of plasticine.) This block of plasticine is 5 m wide.
2. (Draw a pull on the left and an equal pull on the right.) The plasticine is stretched by the forces and gets longer.
3. It was 5 m but has now extended to 12 m. How much has it been extended by?
4. To work out the extension, we subtract the original length from the new length. In this case it has an extension of 7 m. This is the opposite of how we determined the compression.

Make sure students are aware that they can work either out by following the simple rule: *take the small number away from the bigger number.*

Check for understanding

Display the *CFU* slide and click to reveal each question in turn. Using MWBs, ask questions like the ones below.

- If an object that is 8 m wide is squashed to 3 m wide, what is the compression?
- If an object that is 8 m wide is stretched to 10 m wide, what is the extension?
- If a 5 m object is extended by 4 m, how long is the stretched version?
- If a 12 m object is compressed by 7 m, how long is the squashed version?

Independent practice review

Student Practice Book questions 50–55 (pages 117–18).

Q	Notes
55	This draws on prior knowledge. Make sure that students show full working in their answers.

Guided explanation

How do we convert millimetres to metres?

Students always struggle to convert units. Construct the diagram below to help them understand how to convert from millimetres to metres. Add each step sequentially, accompanied by an explanation, similar to the one that follows.

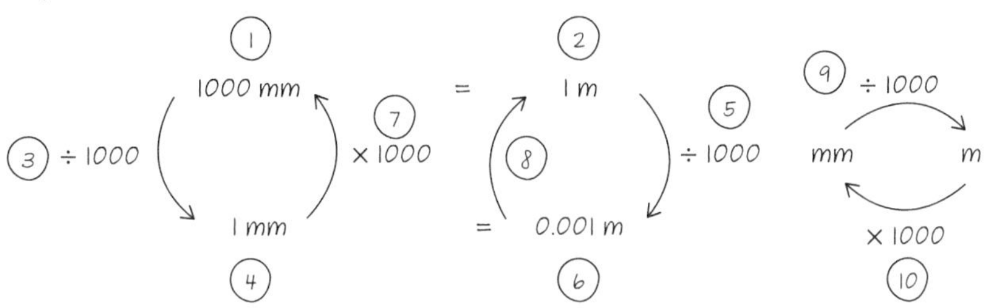

1 On a metre rule (you may like to show one) there are 1000 millimetres.
2 This means that 1000 millimetres is equal to 1 metre.
3 To make it so there is only 1 mm, we need to divide the 1000 mm by 1000.
4 This gives us 1 mm.
5 To calculate how many metres we now have, we need to do the same to the metre.
6 1 divided by 1000 is 0.001 m.
7 If we want to go back from 1 mm to 1000 mm, we need to do the opposite. We multiply by 1000.
8 If we multiply 0.001 by 1000, we get back to 1 m.
9 In summary, to convert millimetres to metres we divide by 1000.
10 And to convert metres to millimetres we multiply by 1000.

Check for understanding

Using MWBs, ask questions like the ones below.

- How many mm are in 1 m?
- How do we convert from mm to m?

- How do we convert from m to mm?
- Display the *CFU* slide and write some values for mm or m. Ask students to convert to the other unit. *Make sure they show their working.*

Independent practice review

Student Practice Book questions 56–63 (pages 118–19).

Q	Notes
56–63	Make sure students show full workings for all answers.

Guided explanation

What is an elastic material?

Students will have an understanding of what elastic materials are, but this needs it to be clearly defined from a scientific perspective.

Construct the diagrams below to ensure that students understand the correct definition of *elasticity*. This is best done with some physical demonstrations, e.g. using plasticine and a spring, or similar. Start by drawing the diagram for squashing, accompanying each step with an explanation, similar to the one that follows.

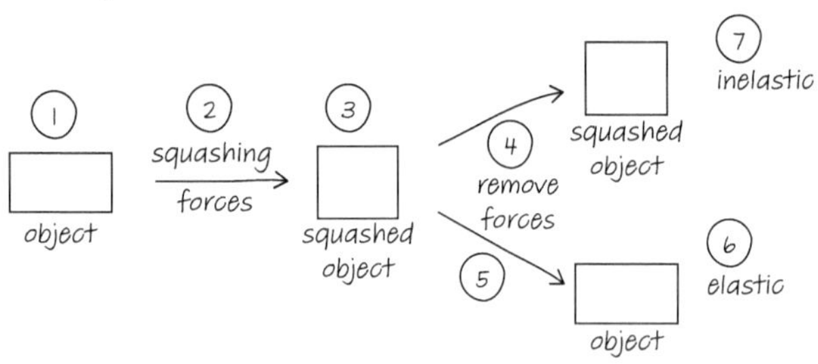

1 Elastic materials have special properties.
2 If an object experiences a squashing force, it becomes squashed.
3 If we remove this force, there are two possible outcomes.
4 One possibility is the object stays in its squashed shape. Plasticine is a good example of this (demonstrate this if you can).
5 The other possibility is that the object returns to its original shape. Springs, sponges and rubber are good examples of this (demonstrate these if you can).
6 Objects that return to their original shape are called *elastic* materials.
7 Objects that stay squashed are not elastic. They are often called *inelastic* materials.

Repeat the same steps with a stretching example, as shown below.

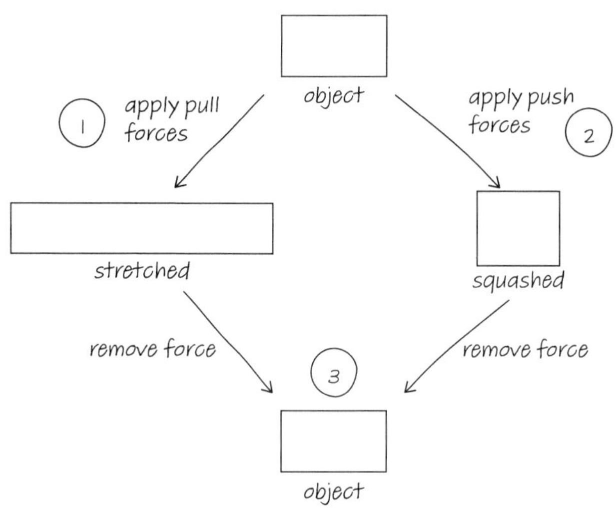

1 Generally, many objects that experience two opposite pull forces will stretch.
2 Or when two opposite push forces are applied, they are squashed.
3 The key thing with elastic objects is that when the forces are removed, the object returns to its original shape.

Check for understanding

Display the *CFU* slide and, using MWBs, ask questions like the ones below.

- What happens to an elastic material when it experiences two pull forces at opposite ends in opposing directions?

- What happens to an elastic material when it experiences two push forces at opposite ends in opposing directions?
- What happens to an elastic material when the forces are removed at opposite ends in opposing directions?
- What can happen to an inelastic material when two push/pull forces are experienced?
- What happens to an inelastic material when the forces are removed?

Independent practice review

Student Practice Book questions 64–76 (pages 119–20).

Q	Notes
67	Needs value and correctly identifying as elastic or not.
70	Needs force and category.
75	Make sure the answer follows the prescribed interactions phrasing to avoid confusion.

P3.4 Hooke's law and work done

Students' learning objective: *I am learning about Hooke's law so I can understand how the force applied affects the extension of a spring.*

Now students have the component skills, we can introduce them to Hooke's law. This will be a good opportunity for them to be introduced to a linear relationship. Note, however, that we will not introduce the terminology of a *proportional* relationship at this stage. Throughout this topic, spend time emphasising this language and the special relationship between the two variables. It can help to have a set of demonstration equipment to show the behaviour of springs and elastic bands, as well as diagrams and drawings, but the practical is best completed after the students are well aware of the law itself.

Knowledge check

After working through the *General science quiz*, display the *Knowledge check* slide and, using MWBs, deliver the questions one by one, reteaching concepts as necessary.

	Question	Answer	Notes
1	What do we call it when two objects affect each other due to forces?	Interaction	
2	If an object experiences two applied forces in opposite directions from either ends, will it be stretched or squashed?	Squashed	
3	What is the name of the non-contact force that pulls you towards the Earth?	Weight	Do not accept gravity.

	Question	Answer	Notes
4	An elastic band is 143 mm long. An applied force stretches it to 233 mm. Calculate the extension.	233 - 143 = 90 mm	Allow conversion to metres.
5	A line of inquiry for this topic is 'increasing the force on a spring increases its extension'. What are the independent and dependent variables?	Independent = the force applied in newtons Dependent = the extension of the spring in metres	Do not penalise students who omit units.

Guided explanation

How does changing the force affect the extension of an elastic object?

This explanation is the beginning of a series that will lead to the introduction of Hooke's law. This section aims to establish the relationship between force and extension. Students should have some real-world experience here, and physical demonstrations of masses with elastic bands and springs will help.

Start with an example of a force being applied to an elastic band. Draw the diagram below sequentially, accompanying each step with an explanation, similar to the one that follows.

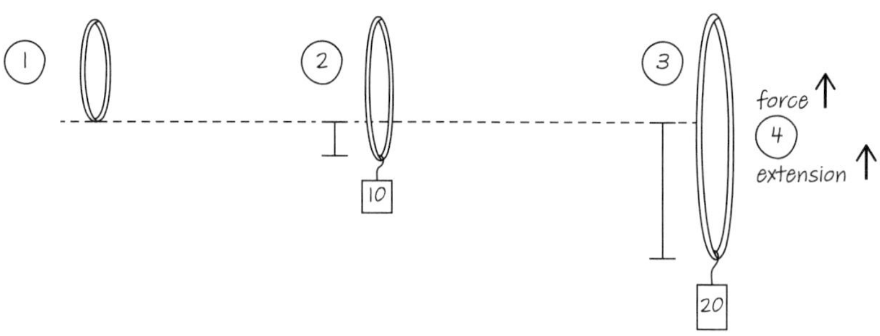

1 Let us look at an example with an elastic band.
2 When we add a force, in this case a weight of 10 N, the band is stretched and has an extension. What happens if we add more force?
3 If we add 20 N of weight, the band will stretch further.
4 As the force increases, the extension increases.

Repeat the same steps for a spring, as shown below.

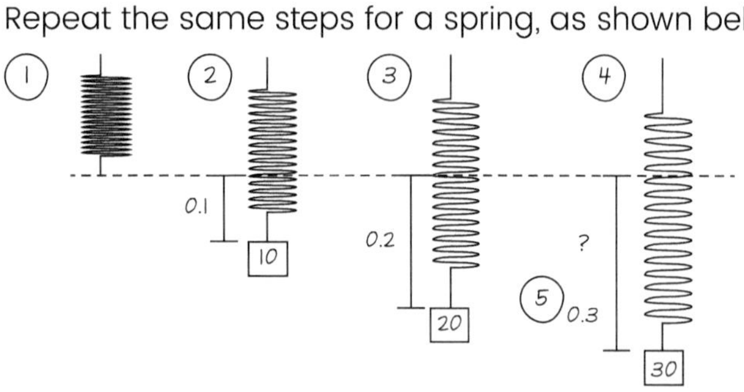

1 The same is true for springs. However, springs are even more special.
2 The extension with a force of 10 N for a given spring is 0.1 m.
3 If we change the weight to 20 N, the extension is 0.2 m.
4 If we add a 30 N weight, what do you predict the extension will be? *0.3 m*
5 When the weight goes up by 10 N, the extension goes up by 0.1 m.

Check for understanding

Using MWBs, ask questions like the ones below.

- When we increase the force on an elastic object, what happens to the extension?
- I have a spring with a large force on it. I take the force off the spring. What happens to the length of the spring?

Display the *CFU* slide and click to reveal each question in turn.

- If a spring extends by 15 cm with 10 N on it, how far will it extend with 20 N on it?
- If a spring with 100 N on it extends by 2.4 m, how far will it extend with 50 N on it?

The independent practice for this section is included in the questions that follow a later guided explanation.

Guided explanation

What is Hooke's law?

Before introducing Hooke's law, we need to explain the spring constant. Draw the diagram below, with each step accompanied by an explanation, similar to the one that follows. Explain that springs are special because they have a relationship between force and extension, but not all springs are as 'springy' as each other.

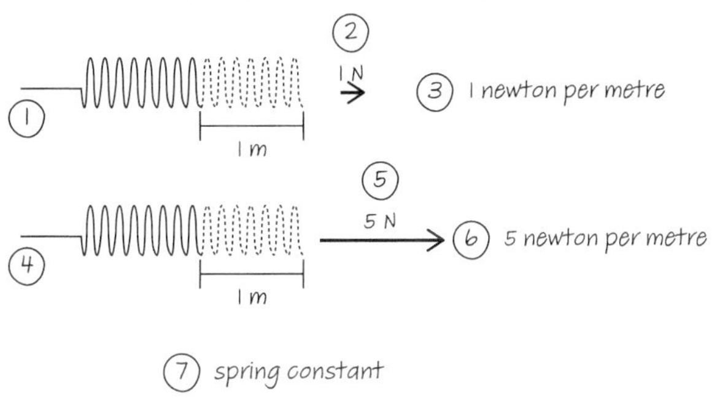

1 Draw the first spring and tell students that it is quite easy to stretch.
2 If we apply a force of 1 N to the spring, it extends by 1 m.
3 We can say that this spring needs a force of 1 N per metre.
4 Draw a second spring and say that it is harder to stretch.
5 To get this spring to stretch 1 m we need a greater force. We need 5 N of force.
6 We can say this spring needs a force of 5 N per metre of extension.
7 The size of the force needed to extend a spring by 1 metre is called the *spring constant*, and it is different for each type of spring.

To establish the relationship between the spring constant, force and extension before deriving the equation, go through the following further examples. Use questioning or MWBs to check student understanding before moving on.

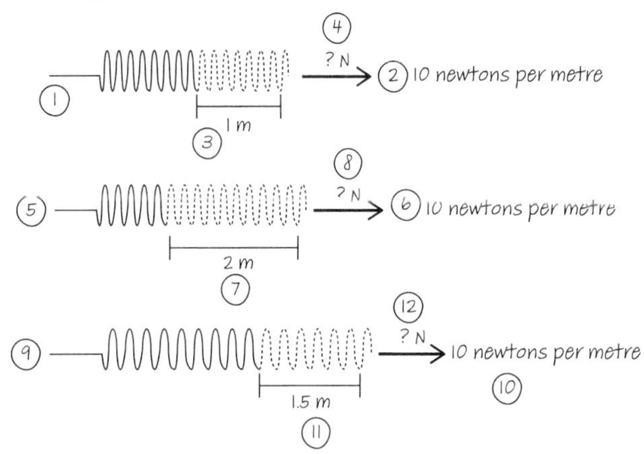

1 Draw a spring similar to before and say that it is a stiffer spring than the last two.
2 It has a spring constant of 10 N/m.
3 It will still extend by 1 metre.
4 How much force is needed? *(10 N)* How did we work that out? *(10 × 1)*
5 Draw the same spring again.
6 It has the same spring constant.
7 This time it is extended 2 m.
8 What force is needed? How did you work that out?
9 Finally, draw the same spring a third time.
10 It has the same spring constant.
11 This time it is extended 1.5 m.
12 What force is needed? How did you work that out?

Tell students that the relationship we have just discovered was first found by a man called Robert Hooke. Demonstrate the equation, as shown below.

(1) 20 N = 10 N/m × 2 m

(2) force = spring constant × extension
(N) (N/m) (m)

(3) F = k × e

1 Hooke noticed what we noticed – that the force (20 N) could be calculated by multiplying the spring constant by the extension.

2 He made this into an equation called *Hooke's law*. It states that the force in newtons is equal to the spring constant in newtons per metre multiplied by the extension in metres.

3 In symbols we write this as $F = k \times e$.

Check for understanding

Using MWBs, ask questions like the ones below.

- What part of Hooke's law describes how stretchy a spring is?
- If a spring has a large spring constant, how easy is it to stretch?
- Display the *CFU* slide and draw two springs, labelled A and B. Ask: If spring A has a larger spring constant than spring B, which one will stretch more when the same force is added?
- If I want spring A to stretch as much as spring B, what do I need to do to the force?
- What is the relationship between force and extension for a particular spring?
- In words, write out the equation that links force, spring constant and extension, often called Hooke's law.
- In symbols, write Hooke's law.
- What is the unit for spring constant/force/extension?

Before moving on to using the equation, ensure that students fully understand the equation in a qualitative sense. The independent practice for this section is included in the questions that follow the guided explanation below.

Guided explanation

How do we calculate the force needed to extend a spring?

Now that students are aware of Hooke's law, we can begin to use it. Students should be aware of the EVERY method from prior units. If this is the first time, ensure that students know why the EVERY method is preferred – to slow them down and prevent silly mistakes in assessments.

Demonstrate how we use Hooke's law to calculate the force we need to apply to get a certain extension, as shown below.

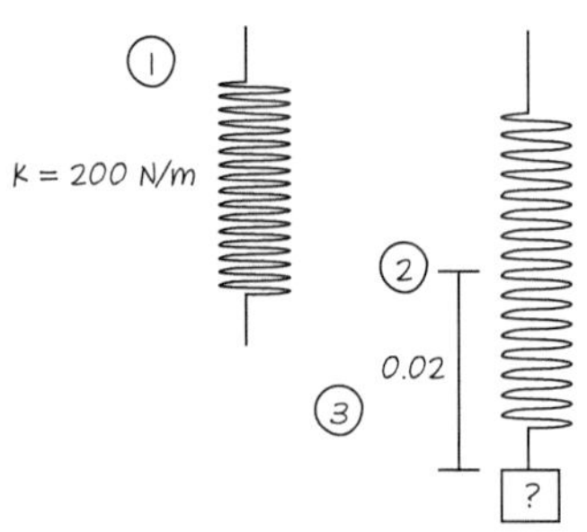

(4) E: F = k × e

(5) V: F = ? k = 200 N/m e = 0.02 m

(6) E: F = 200 × 0.02

(7) R: F = 4

(8) Y: F = 4 N

1 This spring has a spring constant of 200 N/m.

2 A force is added and the spring extends.

3 The extension is measured at 0.02 m.

4 To calculate the force, we use the EVERY method. For E, we write the *equation*.

5 For V, we write the *values*.

6 For E, we then *enter* the values.

7 At R, we get the *result*.

8 And for Y, we add the *units*. From this we can see that Hooke's law tells us to get an extension of 0.02 m we need 4 N of force.

Check for understanding

Display the first two *CFU* slides in turn and work through the examples with the students. Using MWBs, ask questions like the ones below.

- What is the symbol equation that links force, spring constant and extension?
- What is the unit of extension?
- What is the unit of spring constant?

Display each of the remaining slides in turn, with further examples for students to calculate the force needed to extend a spring. Ask questions like the ones below.

- For Example 3, ensure that students have all the steps of the working out, including the correct units.
- Example 4 will require students to use a calculator.
- For Example 5, ensure that students have calculated the extension before calculating the force.

Do not allow students to progress to the independent practice until they are getting questions like these correct.

Independent practice review

Student Practice Book questions 77–90 (pages 120–21).

Q	Notes
81–88	All answers must include the EVERY method to score marks.
89	Credit any answer that links the idea that increasing the spring constant makes a spring that is less stretchy.

Core practical

Investigating the extension of a spring

A method is provided on the *Practical worksheet* and *Teacher and technician notes* on Boost – including equipment list and safety notes – if you wish to allow your students to proceed independently. However, for most students it will ordinarily be better to take a *Slow Practical* approach following the guidance below.

Students should work in groups of two or three. For each group and yourself, you will need eye protection, spring, mass hanger, four 100 g slotted masses, clamp, boss and stand, G-clamp and metre rule.

1. Assign the equipment by getting students to collect apparatus as needed and return to their workstations.
2. Display the first slide and set up a demo station at the front. Show students how to set up the equipment one step at at time.
3. When the students arrive at the step of aligning the metre rule, circulate quickly to ensure that they have all it set up correctly.
4. Pause and introduce the language of errors, which will be a *working scientifically* emphasis. If we read one length incorrectly by mistake we are making a random error. We can try to remove random errors by reading the metre rule at eye level, and always measuring from the same part of the spring.

5 Demonstrate how to safely load the spring, record the extension and unload the spring.
6 Ask students to take and record their first measurement on the *Practical worksheet*. Carefully observe for any issues and then repeat the process. Display the second slide, which shows a template for the results table.
7 If you want students to do repeat measurements, they could then do this independently at their own speed.
8 Ask students to complete the independent practice questions on the *Practical worksheet*. These include plotting a graph. If students have not studied Unit B2 yet, they might find this challenging, so it may need to be carefully modelled under a visualiser. It is recommended that you stop the class after the they have drawn the graph and gain their attention.

Reflecting on the plotted graph, we now need to introduce the idea of a linear relationship.

Display the third slide, showing the graph students plotted for question 8 Add labels sequentially, as shown below, with each step accompanied by an explanation, similar to the one that follows.

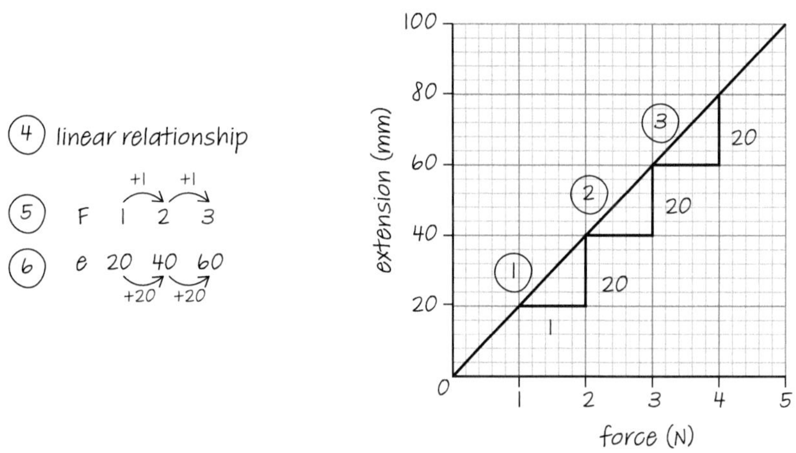

1 Notice that our graphs all produce a nice straight line. This is because when the force goes up by 1 N, the extension goes up by 20 mm.
2 This is repeated.
3 Again and again.
4 The scientific name for a special relationship in which increasing one thing by a set amount creates an increase in another thing by a set amount, is a *linear relationship*.
5 With springs, there is a linear relationship between the weight and the extension. The increase does not have to be the same number. In our example, if the force goes up by 1 N ...
6 ... the extension increases by 20 mm not 1 mm. A linear relationship is saying that every time we add 1 N to the force, we get an increase of 20 mm in the extension. A linear relationship always gives a straight line on the graph.

To check for understanding, display each of the remaining slides in turn, which show a series of graphs. Using MWBs, ask students whether each graph shows a linear relationship. Note that any straight line will be a linear relationship, regardless of where the line intercepts the *x*-axis or if it has a positive or negative gradient. Any curve will not be a linear relationship. A horizontal or vertical line that shows no change in either *x* or *y* is certainly linear, but there is no relationship.

Guided explanation

What kinds of error do we see in experiments?

This explanation introduces the two types of error that students will come across.

What are random errors?

Explain that in experiments, mistakes happen. We call the mistakes *errors*. Display the table on the first slide and ask students to look carefully for mistakes. Give them time and then call their attention back to you. Work through the example using the steps given below.

Force (N)	Student 1 extension (mm)	Student 2 extension (mm)
0	0	0
10	200	250 ②
20	400	400
30	600	600
40	① 700	800
50	1000	1000

③ error too high or low = random error

1 There are two mistakes in this table. The first mistake is 700 mm. If the spring is following Hooke's law, this reading should be 800 mm. The value is lower than we expect. Maybe the person misread the ruler or only added one extra mass when they should have added two.
2 The second mistake is 250 mm. This number is higher than expected. We would expect 200 mm. Maybe the person added too much mass this time or looked at the ruler at an odd angle.
3 Errors are found when a number is too high or too low. These errors are not in a pattern – one is too high the other is too low. We call these types of error *random errors*. They are often caused by humans, and can occur at any time.

What are systematic errors?

Repeat the process with the table on the second slide, as shown below. Explain that a different pair of students did exactly the same experiment with the same spring and obtained this set of results. Ask students to look at the table quietly and then call their attention back to you. Point out that the error in this set of data is different. If it were a random error, it would be both up and down and in unpredictable places. This error is not like that.

Force (N)	Student 1 extension (mm)	Student 2 extension (mm)
0 ①	150 →	0
10 ②	350 →	200
20 ③	550 →	400
30	750	600
40	950	800
50	1150	1000

④ All results too low or too high = systematic error

1 In this experiment, student 1 has included the length of the unstretched spring in their results.
2 Each result is 150 mm longer than it should be (you may want to explicitly show that in both data sets the extensions increase by 200 mm each time).
3 Every time, if we subtract 150 mm from student 1's result, we get student 2's correct result.
4 This type of error, where the same mistake has been made through the whole experiment, leads to all the numbers being too high or too low. We call this a *systematic error*. Systematic errors still display the trend in the results and can often be easily fixed once you spot them. In this case, we could just subtract 150 mm from each extension.

Check for understanding

Using MWBs, ask questions like the ones below.

- What do we call mistakes made in science experiments?
- What types of error are those that make a single result too high or low?
- What types of error are those that make all results too high or low?

Display each of the *CFU* slides in turn and, for each one, ask if it contains random or systematic errors. Make sure to annotate the tables to show where the errors are and what the numbers should have been. In later slides, the error involves the conversion from mm to m, so make sure students can notice this. Hints might be needed for some classes.

Independent practice review

Student Practice Book questions 91–99 (pages 121–23).

Q	Notes
95	*Because-But-So* questions always generate a range of possibly correct answers. You should ask for students with different answers from the mark scheme to read their answers out and discuss whether they are valid.

Guided explanation

What is the elastic limit?

Remind students that springs are elastic, and Hooke's law says that they obey a linear relationship. Draw the diagram below sequentially, with each step accompanied by an explanation, similar to the one that follows.

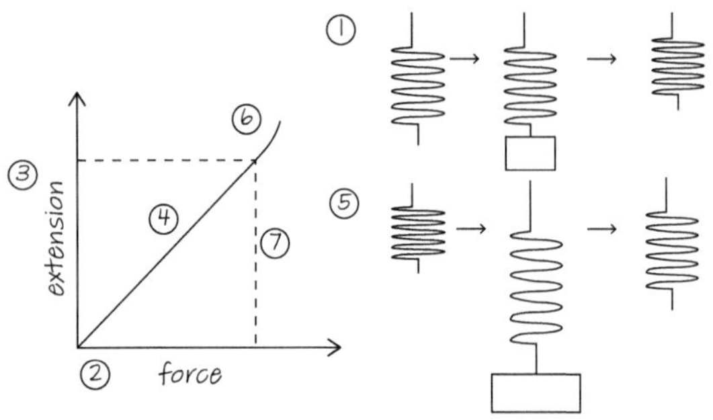

1 When we add a force to the spring, it extends. When we remove the force, it returns to its original size. If a bigger force is added, the spring will extend more and then return to its original size.

2 We can plot this on a graph with force on the *x*-axis ...

3 ... and extension on the *y*-axis.

4 The graph is a straight line. This is why it is called a *linear relationship*. Linear means line.

5 However, this does not go on forever. If a really large force is added to the spring, it gets stretched beyond its ability to return to its original size. The spring is permanently stretched by this force because it is too large. The spring is not elastic for this large force and Hooke's law is broken.

6 On the graph, the line will stop being straight. It is no longer a linear relationship.

7 The point on the line when its stops being straight and Hooke's law breaks down is called the *elastic limit*. All elastic materials that obey Hooke's law have an elastic limit when Hooke's law does not apply any more. Different springs will have different limits, but they all have them.

Check for understanding

Using MWBs, ask questions like the ones below.

- What word do we use to describe a material that can be stretched by a force and then will return to its original size?
- What word do we use to describe a relationship where a regular increase in one thing causes a regular increase in another?
- On a graph, what type of line shows a linear relationship?

Display the *CFU* slide and ask questions like the ones below.

- Which label is the non-linear part?
- Which label is the section that obeys Hooke's law?
- Which label is the part where the spring behaves in an elastic way?
- Which label is the part where the force was too big for Hooke's law to be obeyed.?
- Which label is the part which is beyond the elastic limit?

Independent practice review

Student Practice Book questions 100–05 (page 123).

Q	Notes
104	Make sure students use the EVERY method.
105a	Accept extension in cm or m.

Guided explanation

What is the work done on an elastic material?

Explain that scientists use the phrase *work done* to mean the energy transferred between stores. When we load an elastic material, work is done because energy is transferred between stores. Draw the diagram below sequentially, with each step accompanied by an explanation, similar to the one that follows.

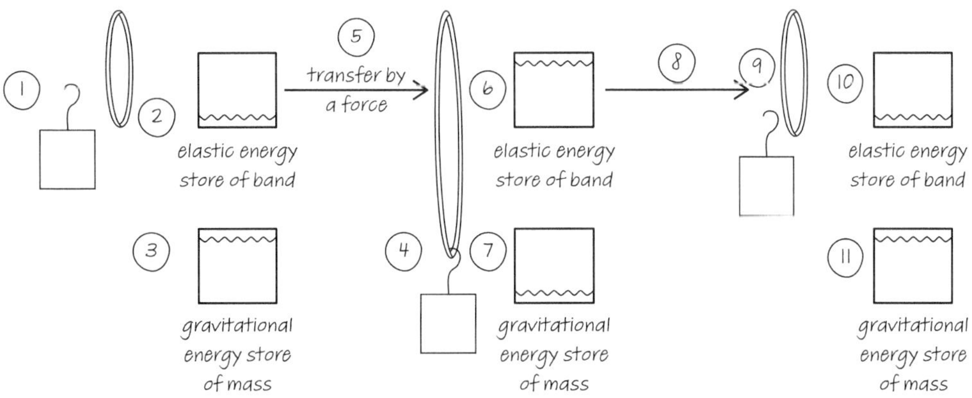

1. An elastic band has a mass hanging next to it.
2. Because the elastic band is not currently stretched, its elastic energy store is empty.
3. The mass is off the ground, so has a lot of energy in its gravitational energy store.
4. When the mass hangs on the elastic band, the weight pulls on the mass and the elastic band stretches.
5. The weight is a force, so the energy transfer is transfer by a force.
6. The band is now stretched, so its elastic energy store has increased.
7. The mass is now nearer the ground, so its gravitational energy store has decreased a bit. We say the elastic band has had work done on it by the mass.
8. When we remove the mass, energy is transferred by a force in the opposite direction. Work is done on the mass by the elastic band.
9. The band contracts.
10. The elastic energy store of the elastic band decreases ...
11. ... and the gravitational energy store of the masses increases again.

Explain that every time an elastic object is stretched, work is done on it by a force. Show students a different example that shows the same concept, but using compression of a spring, as shown below.

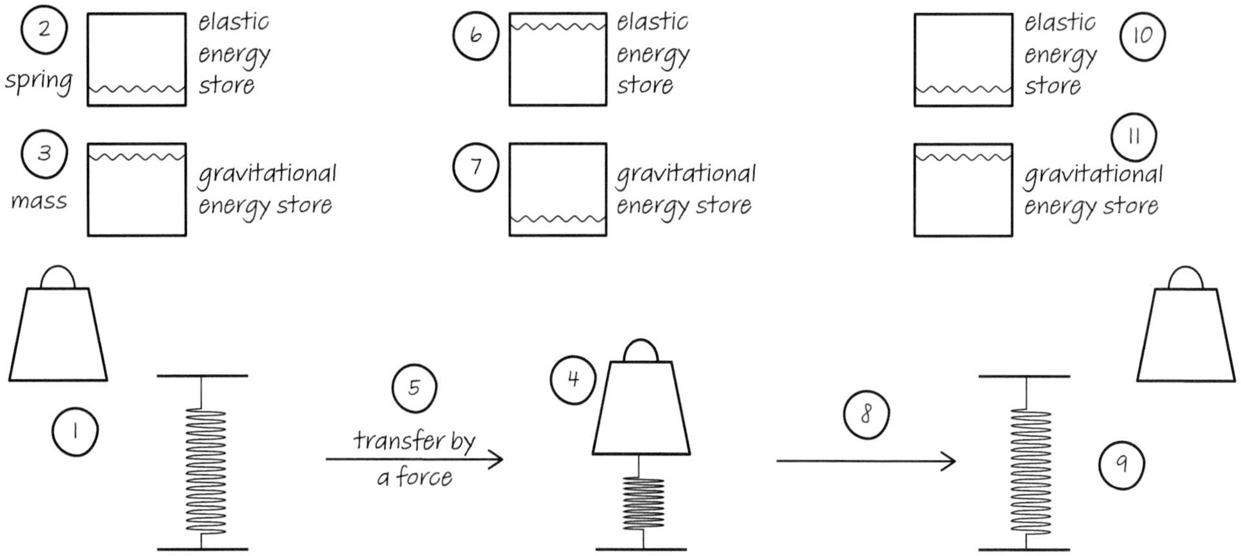

1 A mass is hanging near a spring that is uncompressed.
2 The elastic energy store of the spring is empty.
3 The gravitational energy store of the mass is very high.
4 When the mass is placed on the spring, it compresses the spring. The mass does work on the spring.
5 So energy is transferred by the force of weight.
6 The compressed spring has increased its elastic energy store.
7 The mass is nearer the ground, so its gravitational energy store has decreased. The mass has done work on the spring.
8 As before, the process can be reversed and then the energy can be transferred by the elastic force.
9 The mass is now off the spring. The spring has done work on the mass.
10 So the spring's elastic energy store is empty ...
11 ... and the mass's gravitational energy store is high.

Check for understanding

Display the *CFU* slide and, using MWBs, ask questions like the ones below.

- Which energy store increases when an elastic band is stretched?
- Which energy store decreases when a mass is lowered?
- Which energy store increases when a spring is compressed?
- Which energy store decreases when a spring relaxes?
- What transfers the energy between the elastic energy store and the gravitational energy store?
- What word do scientists use to describe these types of energy transfer?

Independent practice review

Student Practice Book questions 106–13 (pages 124–25).

Q	Notes
106c	Students must identify the gravitational energy store not just the gravitational store.
110	Make sure answers follow the agreed structure, to ensure students can self-mark accurately.
113	Requires knowledge from previous topics. Students should be encouraged to look back at the Knowledge Book if they are stuck.

P3.5 Moments and simple machines

Students' learning objective: *I am learning about moments so I can explain how simple machines work.*

This topic will introduce the concept of moments. Students will have a good understanding of forces' ability to affect objects either vertically or horizontally. Moments will introduce the idea of turning forces and the various applications they provide. School equipment for the moments practical will vary, so please check the equipment your school provides and adjust the method accordingly. Simple machines are covered in this topic, which bridges the gap between the National Curriculum topics of energy changes and forces.

Knowledge check

After working through the *General science quiz*, display the *Knowledge check* slide and, using MWBs, deliver the questions one by one, reteaching concepts as necessary.

	Question	Answer	Notes
1	What is the unit of force?	Newtons	N is acceptable.
2	What is the unit of distance?	Metres	Allow m, but challenge any non-SI unit.
3	What force pulls objects to the centre of the Earth?	Weight	Do not allow gravity.

Guided explanation

What is a moment?

To introduce the concept of moments, it is best to start with a door. This is very familiar to students and readily available for demonstrations.

Show students the classroom door and give them a moment to consider why door hinges are always the opposite side from door handles. Then call their attention back to you and display the first slide. Add to the diagram sequentially, as shown below, with each step accompanied by an explanation, similar to the one that follows.

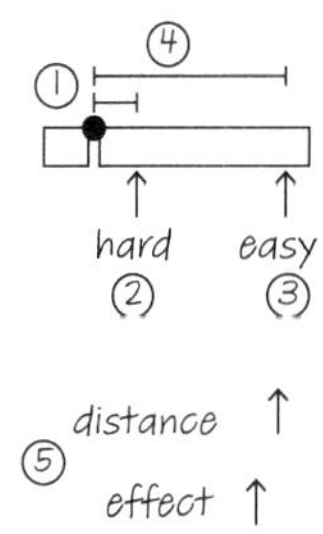

1 Draw the bird's-eye view of the door frame and make sure students can understand what the frame, hinge and door are.
2 Add a push force near the hinges and explain it is hard to open the door. Demonstrate this with classroom door. If the door is particularly light, students will not see how much harder it is to open it if you are pushing close to the hinges. To rectify this, hold the door handle gently and ask them to push the door open. If they push close to the handle, they should be able to do this easily. Closer to the hinge, and they will struggle.
3 Add a push force further away from the hinges with the same force. Ask students how we can tell from the diagram that it is the same force. Explain that it is much easier to open the door.
4 Point out that the only change is the distance from the hinges.
5 Explain that as distance increases, the effect of the force increases.

Tell students that this is not the only thing that can change the effect of the force. Display the second slide and add to the diagram, as shown below, accompanied by an explanation, similar to the one that follows.

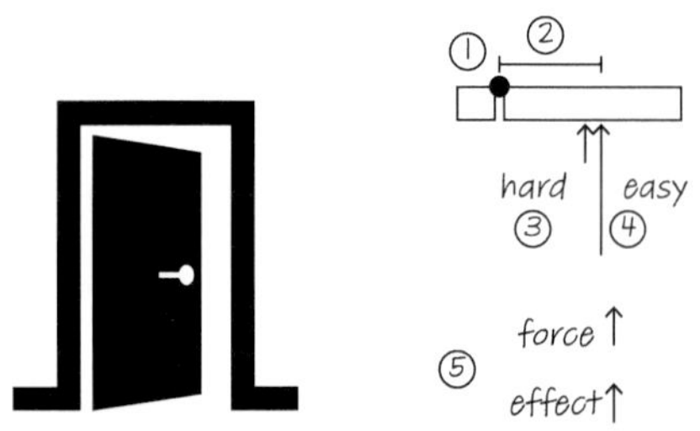

1 Redraw the bird's-eye view of the door.
2 Draw a line to show a constant distance from the hinge.
3 Add a small force causing a small effect and say it is hard to push the door open.
4 Add a bigger force and say this is much easier. The effect has increased.
5 Explain that, in general, we can say that if we increase the force, the effect increases.

Tell students that this works for more things than just doors. Draw the generalised diagram below sequentially, with each step accompanied by an explanation, similar to the one that follows.

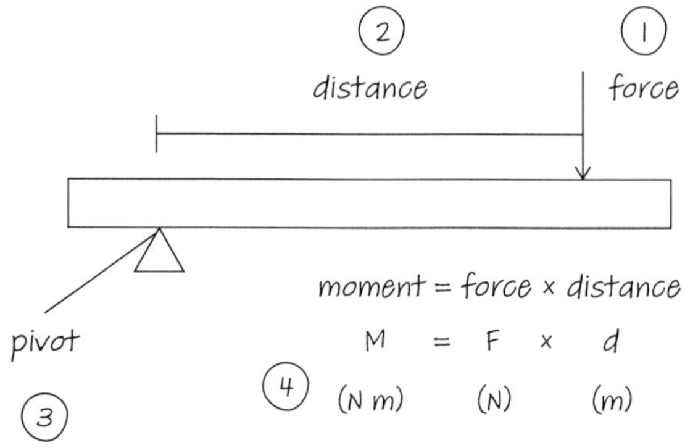

1 Explain that for any situation where there is a force acting ...
2 ... at a distance from a hinge, ...
3 ... or as we call it more generally a *pivot*, there is a turning force.
4 This turning force is given a scientific name because the relationship holds true in all examples. A turning force is called a *moment*. Sometimes you will hear it called the *moment of the force*. The moment can be calculated by multiplying the force in newtons by the distance from the pivot in metres. The unit of a moment is the newtonmetre.

Note that we are not attempting to derive this equation as we have done for others (like Hooke's law). This is because it is harder to do without a stronger conceptual understanding. However, students who understand the principles, relationships and proportions well could be able to understand that *if increasing either of these variables increases the moment, the equation has to have them multiplied together.*

Check for understanding

Display the *CFU* slide and, using MWBs, ask questions like the ones below.

- Which letter is the pivot?
- Which letter is the force?

- What do we call a turning force?
- If the distance from the pivot is increased, what happens to the size of the moment?
- If the force is increased, what happens to the moment?
- What is the word equation that links moment, force and distance?
- What is the symbol equation that links moment, force and distance?
- What are the units of moments?

Independent practice review

Student Practice Book questions 114–22 (page 125).

Q	Notes
118	It is vital that students' terminology is correct here, to avoid issues later in the topic.

Guided explanation

How do we calculate moments?

When showing students how to calculate moments with both the symbol and word equations, make sure they use the EVERY method.

Demonstrate how to calculate moments by drawing the diagram below sequentially, accompanying each step with an explanation, similar to the one that follows.

(1) 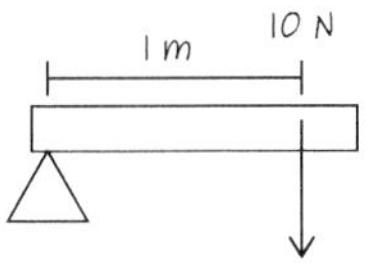

(2) 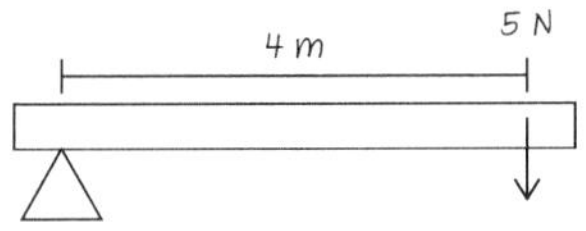

(3) E moment = force × distance
(4) V force = 10 distance = 1
(5) E moment = 10 × 1
(6) R moment = 10
(7) Y moment = 10 N m

(8) E M = F × d
(9) V M = ? F = 5 d = 4
(10) E M = 5 × 4
(11) R M = 20
(12) Y M = 20 N m

1 Draw a simple moments diagram and label the force and distance.
2 Then draw a second example that will give a different moment. Ask the students to look carefully at them and decide which one will have the larger moment.
3 For the first example, write the word equation (E).
4 Write the values (V).
5 Enter the values (E).
6 Calculate the result (R).
7 Add the unit (Y).
8–12 Repeat the steps for the second example.

Tell students that by using the moment equation, we can easily determine the moment of each example.

Check for understanding

Display the *CFU* slide and, using MWBs, ask questions like the ones below.

- Draw a simple moments diagram and label some values. Ask students to complete the calculation step-by-step using the EVERY method. Ask them to show you each line as a class before moving on.
- Repeat this process with another example, increasing their independence once the students have shown they are ready.

Independent practice review

Student Practice Book questions 123–31 (pages 125–26).

Q	Notes
126–31	Make sure students show the EVERY method for the required calculations.

Guided explanation

What are balanced moments?

Now that students can successfully calculate an individual moment, they can be introduced to balanced moments.

Demonstrate balanced moments with the following examples, using the EVERY method.

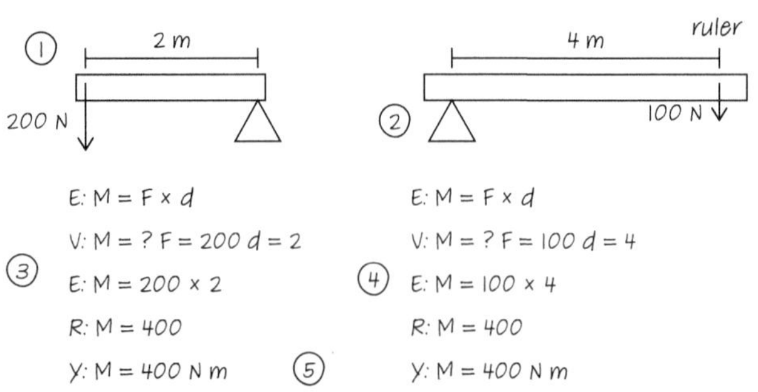

1. Draw a moment diagram, labelled with a force of 200 N and a distance of 2 m.
2. Draw a second diagram with a longer arm, labelled with a force of 100 N and a distance of 4 m. Ask students which moment is greater.
3. Calculate the moment on the left. *400 Nm*
4. Calculate the moment on the right. *400 Nm*
5. Point out that even though the forces and distances are different, the moments are equal.

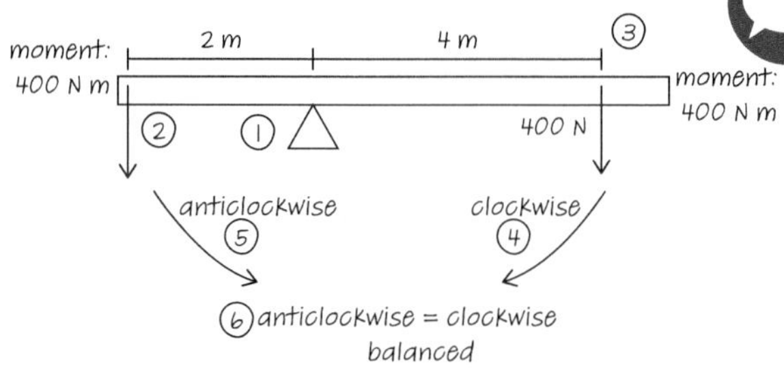

1. Draw a see-saw and explain that it was made by putting together the two from the previous example, so 2 m is on one side of the pivot and 4 m on the other.
2. The moment on the left was calculated to be 400 Nm.
3. The moment on the other side was also 400 Nm.
4. The right-hand side moment is known as the *clockwise moment*, because if that side starts to fall, the plank will spin clockwise.
5. The left-hand side moment will move in the opposite direction, so is known as the *anticlockwise moment*.
6. When the clockwise moment equals the anticlockwise moment, we say the moments balance.

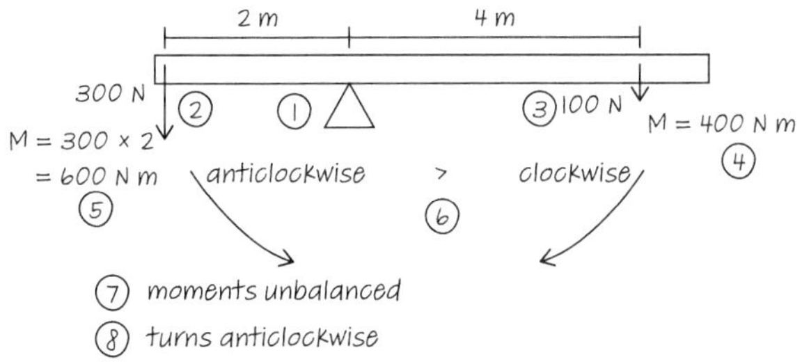

1 Explain that moments do not always balance.

2 Draw a greater force on the left of the pivot.
3 Keep the force on the right the same.
4 The right side still has a moment of 400 Nm.
5 The left side now has a moment now of 600 Nm.
6 The anticlockwise moment is larger than the clockwise moment.
7 The moments are unbalanced.
8 When the moments are unbalanced, the see-saw must move. In this case, it moves anticlockwise.

Check for understanding

Using MWBs, ask questions like the ones below.

- What are the two directions a moment can turn?
- What is special about moments when they are balanced?

Display the *CFU* slide and add simple numbers for forces and distances to the see-saw. Leave out one value and ask students to write this missing value.

Independent practice review

Student Practice Book questions 132–33 (page 126).

Q	Notes
132	Students should be encouraged to show their workings to prove if the see-saw is balanced if they are finding it hard to complete the task.

Core practical

Investigating balanced moments

A method is provided on the *Practical worksheet* and *Teacher and technician notes* on Boost – including equipment list and safety notes – if you wish to allow your students to proceed independently. However, for most students it will ordinarily be better to take a *Slow Practical* approach following the guidance below.

Students should work in groups of two or three. For each group and yourself, you will need a metre rule, prism or other triangular pivot and ten 100 g slotted masses.

1 Collect a set of equipment at the front and show the class each piece.
2 Assign the equipment by getting students to collect apparatus as needed and return to their workstations. Set up a demo station at the front.
3 Ensure that students are aware of the force each mass will generate. This will most likely be 1 N.
4 Demonstrate how to set up the see-saw.
5 Place one mass on either side at equal distances, so the see-saw balances. (Display the first slide, which shows the set up.)
6 Instruct the students to do the same and monitor them. You should be able to see any issues because the see-saws will not balance.
7 Demonstrate the balancing of a single mass and a double mass. Ask students clarifying questions about the size of the force and distance.
8 Display the second slide and show students how to record their results in the table.
9 Get students to collect and record their data on the *Practical worksheet*.
10 At this stage you should feel that students are comfortable to complete the rest on their own, and to choose their own combinations.
11 Ask students to complete the independent practice questions on the *Practical worksheet*.

Guided explanation

What is an anomalous result?

Explain that an anomalous result is a result that does not fit the trend (pattern). These are often the result of random errors that occur during an experiment. We can identify anomalies by looking at the trend. Work through the following example of a series of data. Display the slide and add notes as shown, each accompanied by an explanation, similar to the one that follows.

12, 13, 11, 14, 11, 13, 12 ① no anomalies

② 12, 13, (19), 11, 12, 14, 13 too high

③ 10, 20, 19, 14, 16, 17, 12 not anomalous

④ (6), 20, 19, 22, 18, 17, 21 too low

⑤ anomalous = does not fit trend

1 In the first set of data, all the numbers are close together and within a range of 11–14. There are no anomalies in this set of results.
2 In the second set of results, there is an anomaly. The value 19 is much higher than the rest, so it is an error, or anomaly.
3 In the third set of results, the value of 19 is not an anomaly because it is within the range of the answers, 10–20. Anomalies are always far away from the rest of the answers.
4 The final set of data has an anomaly that is too small. The value 6 is an anomalous result because it is much lower than the rest of the results, which are in the range of 17–22.
5 Remember that anomalous results are ones that do not fit the trend.

Check for understanding

Display the *CFU* slide and click to reveal each row of numbers in turn. Using MWBs, ask questions like the one below.

- Which is the anomalous result?

The independent practice for this section is included in the questions that follow the guided explanation below.

Guided explanation

What are simple machines?

Students will now apply their understanding of moments to simple machines like levers. It is best at this stage to take a qualitative approach, because rearranging equations has not been covered yet.

Display the first slide. Add to the diagram as shown below, accompanying each step with an explanation, similar to the one that follows.

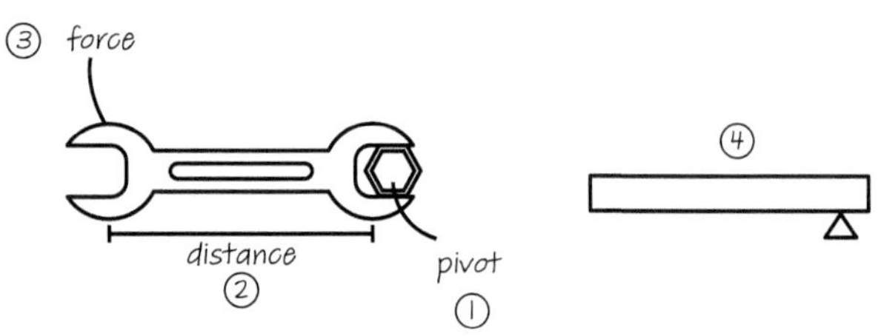

1 The simplest machine is a spanner. The nut needs to turn, so it is a pivot.
2 The spanner's arm provides a distance to create a moment.
3 The force is applied at the other end of the spanner. The moment means we can use less force because the distance increases the moment.
4 Draw a moment diagram as before to show the similarities and then discuss how they can change the spanner to increase the moment by lengthening the arm.

On the second slide, draw the diagram below as an example of a lever, accompanying each step with an explanation, similar to the one that follows.

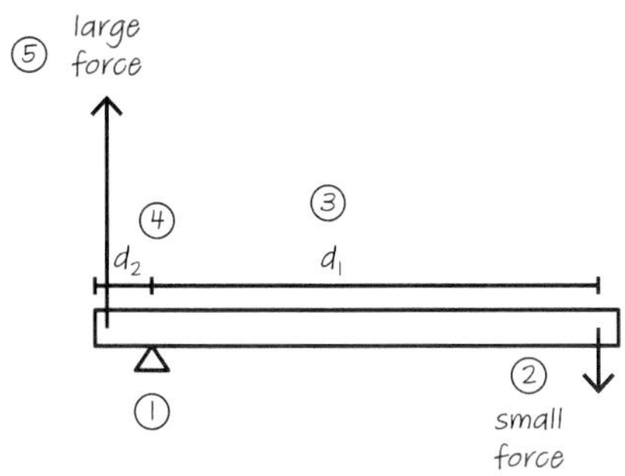

1 Levers work because you put the pivot really close to one side.
2 If you push a small force ...
3 ... a long distance away from the pivot, this creates a large moment.
4 If there is a small distance the other side ...
5 ... the moments must be equal, so the force needs to be much larger. A good example of this is using a screwdriver to open a tin of paint.

Say that wheelbarrows are also levers. They just have both moments on the same side of the pivot. On the third slide, draw the following diagram, accompanied by an explanation, similar to the one that follows.

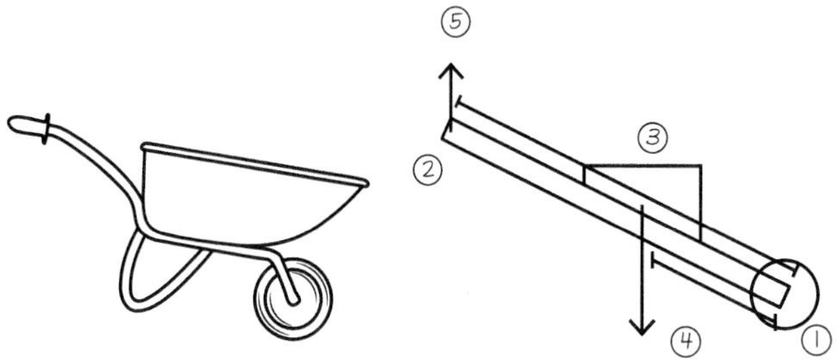

1 On a wheelbarrow, the wheel is the pivot.
2 There is a long arm, so the handles are far from the pivot.
3 The load being carried is in the middle of the wheelbarrow.
4 There is a large force a short distance away from the pivot.
5 Because the arm is long, the distance from the pivot is large, so the force needed to lift the wheelbarrow is lower.

Note that this example is just illustrative. Detailed understanding beyond *the handles of the wheelbarrow are a long way from the pivot (the wheel)* is not demanded at this point.

Display the fourth slide and summarise by saying that the wheelbarrow, lever and spanner are all examples of simple machines. They use the idea of moments to allow us to increase the force we can apply. Other simple machines exist, like pulleys, but we will just focus on these examples.

Check for understanding

Using MWBs, ask questions like the ones below.

- To reduce the force needed, what should we do to the length of a lever?
- Which wheelbarrow is better, one with longer handles or one with shorter handles?

Display the wheelbarrow on the *CFU* slide and ask questions like the ones below.

- Which letter is the pivot?
- Which letter is the load?
- Which letter shows the distance between the pivot and the load?
- Which letters should be multiplied together to find the moment on the handle?
- Why are long spanners better at removing nuts?

Independent practice review

Student Practice Book questions 134–45 (pages 127–29).

Q	Notes
136	Make sure workings are shown for conversion and calculation.
140	*Because-But-So* questions can often generate a range of responses, so it is best to explore the possible answers students wrote and discuss their individual merits.

P3.6 Balanced forces

Students' learning objective: *I am learning about balanced forces so I can explain why an object's speed does not change.*

By now students have a strong grasp of forces that work hortizontally, vertically and in turning situations. This topic allows us to revisit the work on interactions and look at the language of balanced and resultant forces. The focus is mainly on horizontal interactions, but you could extend these explanations to include moments and other examples. This topic will refer to *balanced forces* rather than *equilibrium* to keep the explanations simple to understand.

Knowledge check

After working through the *General science quiz*, display the *Knowledge check* slide and, using MWBs, deliver the questions one by one, reteaching concepts as necessary.

	Question	Answer	Notes
1	How do we represent forces in diagrams?	Arrows	
2	What is the relationship between the size of the force and the length of the force arrow?	The longer the arrow the greater the force	Accept the inverse of this.
3	What do we call a pair of forces that are caused when two objects are acting on each other?	An interaction	
4	If a bottle of water has a weight of 50 N, what is the normal reaction force pushing up from the table?	50 N	

Guided explanation

What are balanced forces?

Introduce the concept of balanced forces with the diagram below, drawn sequentially, with each step accompanied by an explanation, similar to the one that follows.

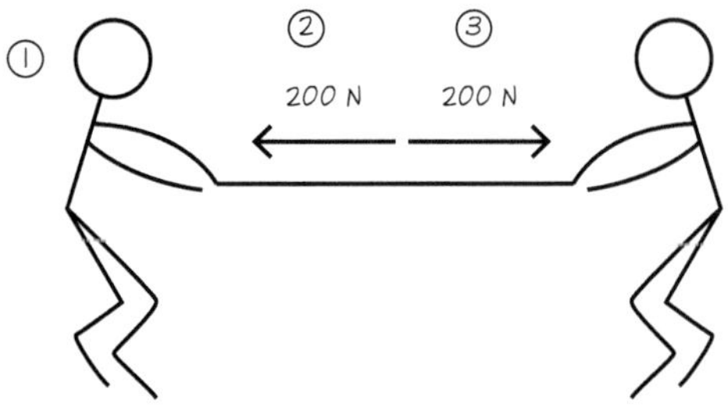

1. Draw two people playing a tug of war.
2. Say that one is pulling with a force of 200 N.
3. The other is also pulling with a force of 200 N, but in the opposite direction. Ask students who will win.

Explain that because the forces are equal in size but opposite in direction, they cancel each other out. We say the forces are *balanced*.

A diagram for this can be found in Topic P3.6 of the Knowledge Book.

Check for understanding

Using MWBs, ask questions like the one below.

- What are balanced forces?

Display each of the *CFU* slides in turn, which show a range of different situations, and ask questions like the ones below.

- Draw the missing force.
- Label the forces.
- Are the forces contact or non-contact?
- Add random values for an individual force and ask students what the opposing force's value would be if they were balanced.

Independent practice review

Student Practice Book questions 146–51 (pages 129–30).

Q	Notes
146	The answer needs the size and direction to score mark.
149	The answer needs the name of the force, the size and the direction.
151c	Students must show use of the EVERY method.

Guided explanation

What is a resultant force?

Explain that forces are not always balanced by drawing the diagram below sequentially, with each step accompanied by an explanation, similar to the one that follows.

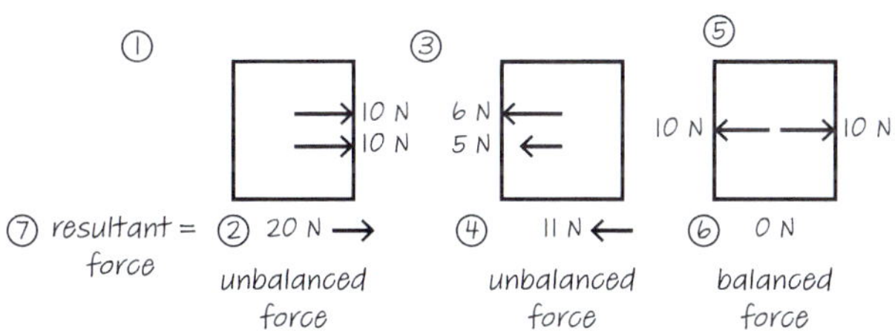

1. This box has two applied forces pushing right of 10 N. It could be that two people are pushing on it. Objects can experience more than one force at a time, and they can be in any direction.
2. These two people work together and create an overall force of 20 N to the right. This is an unbalanced force.
3. The next box has two applied forces to the left, 5 N and 6 N.
4. The overall force is 11 N to the left. (To get the total, we add them together.) This is also an unbalanced force, but acting to the left instead of the right.
5. The final box also has two applied forces, both 10 N, but this time there is one on each side.
6. These forces cancel each other out, so the overall force is zero. These are *balanced forces*, like we have already seen.
7. We have a scientific name for the overall force. We call it the *resultant force*. The resultant force is the overall effect of all the forces on an object. In the last example, the resultant force is zero and the forces are balanced. When the resultant force is zero, we have balanced forces.

Return to the tug of war example with different values for the forces, as shown below.

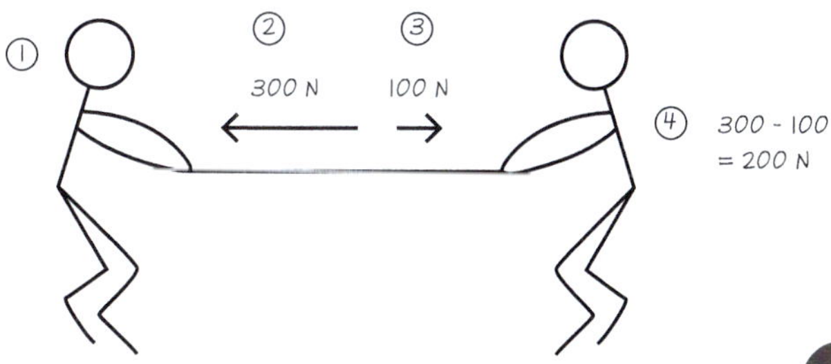

1 One person is now pulling with a force of 300 N.
2 The other person is pulling with a force of 100 N.
3 The forces are not balanced, so there will be a resultant force above zero. The difference between the two forces will tell us the resultant force. In this case, there is a resultant force of 200 N to the left.

Explain that there can be more than two forces involved. In any situation, the overall force is the resultant force.

Check for understanding

Using MWBs, ask questions like the one below.

- What is the scientific word for the resultant force?

Display each of the *CFU* slides in turn and, for each one, ask students to calculate the size and direction of the resultant force. Make sure students show their working as they calculate the resultant force for a mixture of examples. Use additional examples until students are consistently getting them correct.

Independent practice review

Student Practice Book questions 152 and 153 (page 130).

Q	Notes
153	The answers must include size and direction, unless it is 0 N.

Guided explanation

How does a resultant force of zero affect an object?

This is a brief introduction, which will be built on in the next topic. Our aim here is to only discuss the effect of a resultant force of zero.

Draw the diagram below sequentially, with each step accompanied by an explanation, similar to the one that follows.

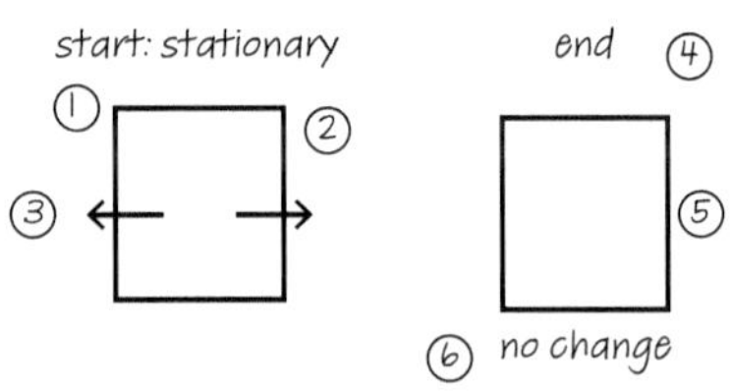

1 Draw a box and say that it is stationary.
2 Draw a push force to the right.
3 Add a balancing push force to the left. Discuss what the resultant force will be. Make sure that students can identify it as zero.
4 Draw the end result.
5 Show that the forces cancel out, so it is like no force is acting.
6 Because there is no resultant force, there is no change.

Explain that when things are moving, it can get a little more complicated. Draw the diagram below sequentially, with each step accompanied by an explanation, similar to the one that follows.

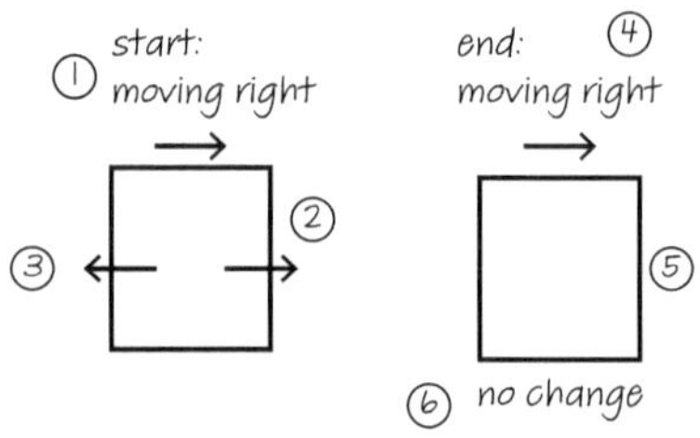

1 Say that this time the box is sliding along the ground, moving right, at a constant speed.
2 A person adds an applied force of 10 N. They push it from behind, trying to speed it up.
3 However, another person adds an applied force of 10 N – they push in the opposite direction. What will happen to the motion of the box?

4 The two applied forces will cancel out. The resultant force will be zero, so there is no change.
5 The box does not stop. It will keep moving at a constant speed.
6 There is no change to the motion.

Tell students this general rule, and write it on the board: If the resultant force is zero, the object does not have its movement changed at all.

Check for understanding

Using MWBs, ask questions like the ones below.

- What does the word stationary mean?
- What happens to a stationary object when the resultant force is zero?
- What happens to a moving object when the resultant force is zero?

Display each of the *CFU* slides in turn and, for each one, ask students what the resultant force will be in each case. When the answer is zero, ask what the final motion will be. Ensure that students use phrases like *it will continue moving to the right at the same speed.*

Independent practice review

Student Practice Book questions 154–63 (pages 130–31).

Q	Notes
154	Allow similar wording that is not identical.
159	The answer must show the EVERY method.

P3.7 Forces and motion

Students' learning objective: *I am learning how resultant forces affect motion so I can explain why objects change speed.*

The final section combines previous content to help explain non-zero resultant forces. Students will have plenty of observations from everyday life to draw on, but often get confused with the idea of non-resultant forces and the effects they have. While the National Curriculum aims at a more qualitative understanding, this course includes a quantitative explanation, because this will reinforce the explanation and aid students' understanding, without significantly increasing the demand mathematically.

Knowledge check

After working through the *General science quiz,* display the *Knowledge check* slide and, using MWBs, deliver the questions one by one, reteaching concepts as necessary.

	Question	Answer	Notes
1	Which has a larger force – a longer arrow or a shorter arrow?	Longer arrow	
2	What is the resultant force?	The overall force	Accept descriptions of the process.
3	What happens to a moving object if the resultant force is zero?	There is no change to its motion	Accept 'it keeps moving at the same speed'.
4	What happens to a stationary object if the resultant force is zero?	It stays stationary	Accept 'no change'.

Guided explanation

What happens to stationary objects when they experience a resultant force?

These diagrams will be similar to those in the last topic, to maintain familiarity. We are using a solid arrowhead to show movement, so the students do not confuse it with a force. You might choose a different way to differentiate, perhaps using colour. It does not matter as long as the students are aware of the difference, and it is consistent.

Draw the diagram below sequentially, with each step accompanied by an explanation, similar to the one that follows.

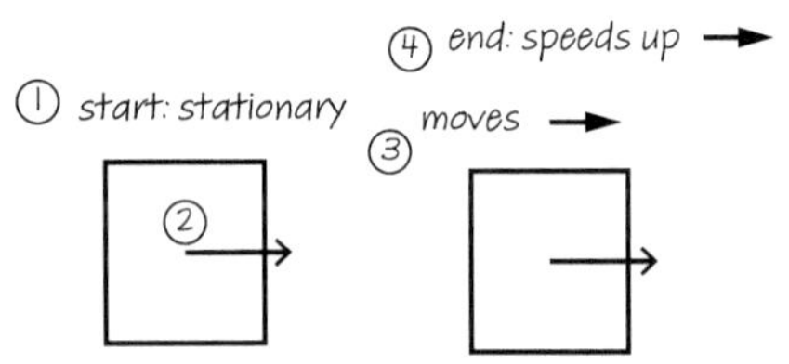

1 Draw a box and say that it is stationary.
2 Add a single force arrow. Discuss the resultant force now it is above zero.
3 Add a label to say that the box is moving.
4 A resultant force above zero causes a stationary object to speed up in the same direction as the force.

Repeat the process with a second example, as shown below.

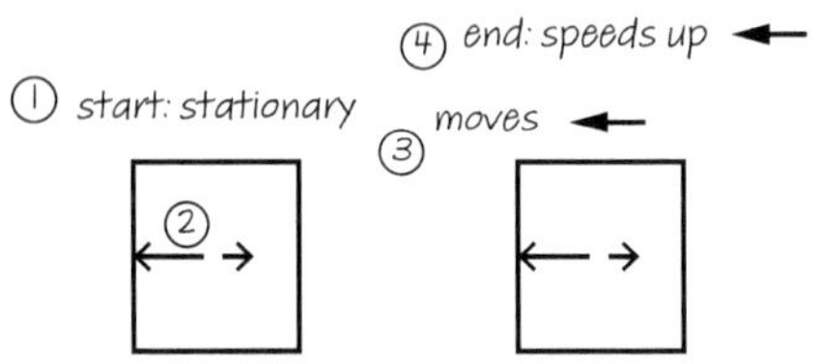

1 Start with the stationary box again.
2 This time add two forces and discuss the direction of the resultant force.
3 Draw the box moving to the left.
4 Say that it speeds up in the direction of the resultant force.

Check for understanding

Display each of the *CFU* slides in turn, which show boxes and different arrangements of forces. Using MWBs, ask questions like the ones below.

- What is the resultant force?
- What will the motion of the box be?

Independent practice review

Student Practice Book questions 164–67 (page 132).

Q	Notes
166	The answers need force, direction and effect.
167b	Ensure that EVERY is used for full marks.

Guided explanation

What happens to moving objects when they experience a resultant force?

This explanation will look at the different scenarios that objects can face due to changing forces. These be introduced one at a time, and the change in speed and direction of each discussed in turn.

1 Display the slide and click to reveal each scenario, one at a time.
2 Give students time to read the table in silence before calling their attention back to you.
3 Describe the starting situation, the resultant force and the outcome of the resultant force.
4 Repeat this for each row.

A copy of the table can be found in Topic P3.7 of the Knowledge Book.

Check for understanding

Display each of the *CFU* slides in turn, which show boxes moving in various directions with different forces. Using MWBs, ask questions like the ones below.

- Calculate the resultant force.
- Describe how the motion will change.

Independent practice review

Student Practice Book questions 168–75 (pages 132–34).

Q	Notes
168	Answers must include size of force, direction of force and change in motion to score the mark.
175c, d	Students must refer to the energy store, not just energy.
175g	Ensure that EVERY is used for full marks.